School Years
2016-2017 through 2020-2021

Written by

Steven R. Conrad • Daniel Flegler • Adam Raichel

Published by MATH LEAGUE PRESS
Printed in the United States of America

Cover art by Bob DeRosa

First Printing, 2021

Math League Press
P.O. Box 17
Tenafly, NJ 07670-0017

ISBN 978-0-940805-25-5

Preface

Math Contests—Grades 7 and 8, Volume 8 is the eigth volume in our series of problem books for grades 7 and 8. The first seven volumes contain the contests given in the school years 1979-1980 through 2015-2016. This volume contains the contests given from 2016-2017 through 2020-2021. (You can use the order form on page 154 to order any of our 24 books.)

This book is divided into three sections for ease of use by students and teachers. You'll find the contests in the first section. Each contest consists of 30 or 35 multiple-choice questions that you can do in 30 minutes. On each 3-page contest, the questions on the 1st page are generally straightforward, those on the 2nd page are moderate in difficulty, and those on the 3rd page are more difficult. In the second section of the book, you'll find detailed solutions to all the contest questions. In the third and final section of the book are the letter answers to each contest. In this section, you'll also find rating scales you can use to rate your performance.

Many people prefer to consult the answer section rather than the solution section when first reviewing a contest. We believe that reworking a problem when you know the answer (but *not* the solution) often leads to increased understanding of problem-solving techniques.

Each school year, we sponsor an Annual 7th Grade Mathematics Contest, an Annual 8th Grade Mathematics Contest, and an Annual Algebra Course 1 Mathematics Contest. A student may participate in the contest on grade level or for any higher grade level. For example, students in grade 7 (or below) may participate in the 8th Grade Contest. *Any* student may participate in the Algebra Course 1 Contest. Starting with the 1991-92 school year, students have been permitted to use calculators on any of our contests.

Steven R. Conrad, Daniel Flegler, & Adam Raichel, contest authors

Acknowledgments

For her continued patience and understanding, special thanks to Marina Conrad, whose only mathematical skill, an important one, is the ability to count the ways.

For demonstrating the meaning of selflessness on a daily basis, special thanks to Grace Flegler.

To Jeannine Kolbush, who did an awesome proofreading job, thanks!

Table Of Contents

The Contests

2016-2017 through 2020-2021

7th Grade Contests

2016-2017 through 2020-2021

Math League Press, P.O. Box 17, Tenafly, New Jersey 07670-0017

2016-2017 Annual 7th Grade Contest

Tuesday, February 21 or 28, 2017

7

Instructions

- **Time** Do *not* open this booklet until you are told by your teacher to begin. You might be *unable* to finish all 35 questions in the 30 minutes allowed.
- **Scores** Please remember that *this is a contest, and not a test*—there is no "passing" or "failing" score. Few students score as high as 28 points (80% correct). Students with half that, 14 points, *should be commended!*
- **Format, Point Value, & Eligibility** Every answer is an A, B, C, or D. Write answers in the *Answers* column. A correct answer is worth 1 point. Unanswered questions get no credit. You **may** use a calculator.

1. The product of -4 and _?_ is 12.
   A) $-\frac{1}{3}$ B) -3 C) 3 D) $\frac{1}{3}$ — 1.

2. The ratio of pens to pencils on my desk is 5:8. There could be _?_ pencils on my desk.
   A) 10 B) 12 C) 14 D) 16 — 2.

3. The product of 5.96 and 3.06, rounded to the nearest whole number, is
   A) 15 B) 16 C) 18 D) 20 — 3.

4. $11 - 22 + 33 - 44 + 55 - 66 + 77 =$ _?_
   A) 33 B) 44 C) 77 D) 110 — 4.

5. Of the following fractions, which has the least value?
   A) $\frac{\sqrt{2}}{\sqrt{3}}$ B) $\frac{\sqrt{3}}{\sqrt{4}}$ C) $\frac{\sqrt{4}}{\sqrt{5}}$ D) $\frac{\sqrt{5}}{\sqrt{6}}$ — 5.

6. Which of the following numbers is not a factor of $4 \times 6 \times 8$?
   A) 16 B) 24 C) 30 D) 96 — 6.

7. The ages of my 3 cousins and my 3 uncles are 14, 15, 16, 42, 45, and 48. Their average age is
   A) 30 B) 29 C) 28 D) 27 — 7.

8. $4 - 2 \div \left(\frac{1}{2}\right) =$
   A) 0 B) 3 C) 3.75 D) 4 — 8.

9. My rectangular field is 80 m long, 40 m wide, and bordered on all sides by a fence. If the same total length of fencing borders a square field, the area of the square is _?_ greater than the area of my field.
   A) 200 m^2 B) 400 m^2 C) 600 m^2 D) 800 m^2 — 9.

10. Maggie has only pennies, nickels, and dimes. If she chooses exactly 8 coins, the amount of money she chooses *cannot* be
    A) 70¢ B) 65¢ C) 35¢ D) 15¢ — 10.

11. $(42 \times 54 \times 58 \times 46) \div (29 \times 21 \times 23 \times 27) = 2 \times$ _?_
    A) 2 B) 4 C) 6 D) 8 — 11.

12. Two numbers are reciprocals. If the first one is multiplied by 4, the second one must be multiplied by _?_ for the two products to be reciprocals.
    A) -4 B) 1 C) $-\frac{1}{4}$ D) $\frac{1}{4}$ — 12.

13. All sides of a _?_ can have different lengths.
    A) parallelogram B) rhombus C) trapezoid D) square — 13.

Go on to the next page ⟹ 7

14. Shayna ran each of the following distances in km:

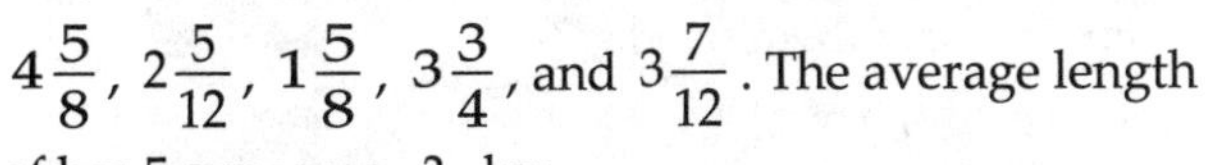

$4\frac{5}{8}$, $2\frac{5}{12}$, $1\frac{5}{8}$, $3\frac{3}{4}$, and $3\frac{7}{12}$. The average length of her 5 runs was _?_ km.

A) $3\frac{1}{4}$ B) $3\frac{1}{3}$ C) $3\frac{1}{5}$ D) $3\frac{1}{2}$

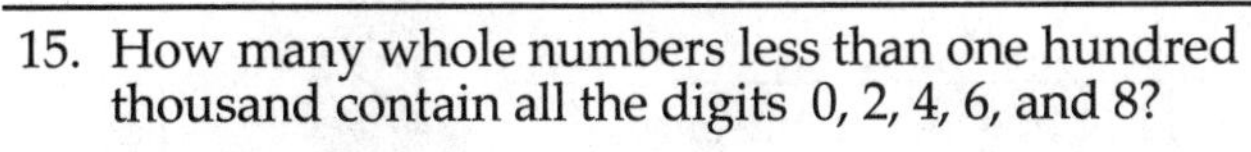

15. How many whole numbers less than one hundred thousand contain all the digits 0, 2, 4, 6, and 8?

A) 625 B) 384 C) 120 D) 96

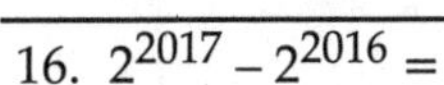

16. $2^{2017} - 2^{2016} =$

A) 1 B) 2 C) 2^{2016} D) 2^{2017}

17. Of the following, which is the largest number?

A) $\frac{-2-3}{-1}$ B) $\frac{12-4}{3}$ C) $\frac{12+3}{5}$ D) $\frac{8-2}{-1}$

18. What is the least number of different single-digit numbers I can choose so that the product of these numbers is greater than 2017?

A) 3 B) 4 C) 6 D) 7

19. Which of the following is not divisible by 7?

A) 7677 B) 7686 C) 7714 D) 7777

20. If there are 8 nits in a nat, how many nits are in 6.25 nats?

A) 49 B) 50 C) 51 D) 52

21. Before her 3 new baby female lizards hatched, 40% of Cecily's lizards were female. Now 50% are female. How many lizards does she have?

A) 12 B) 15 C) 16 D) 18

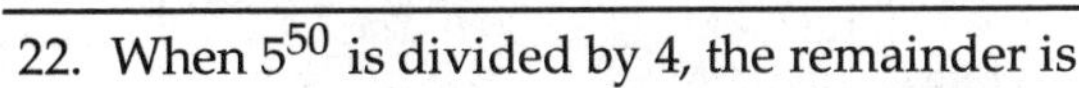

22. When 5^{50} is divided by 4, the remainder is

A) 0 B) 1 C) 2 D) 3

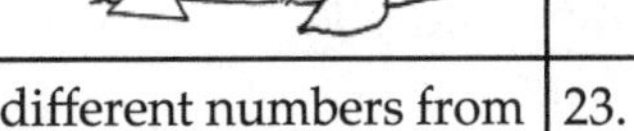

23. In how many different ways can I choose two different numbers from 67, 71, 75, 80, 87 so that their sum is odd?

A) 4 B) 8 C) 24 D) 60

24. If the sum of two whole numbers is 100, what is the least possible value of the sum of the reciprocals of these two numbers?

A) 0.01 B) 0.04 C) 0.05 D) 0.1

25. Which of the following statements is true?

A) $(-0.5)^3 > (-0.5)^2$ B) $(-0.5)^3 < (-0.5)^2$

C) $(-0.5)^3 = (-0.5)^2$ D) $(-0.5)^3 = -(-0.5)^2$

Answers
14.
15.
16.
17.
18.
19.
20.
21.
22.
23.
24.
25.

Go on to the next page ➠

26. At most how many 10 cm-by-13 cm tiles can fit on the floor of a 91 cm-by-130 cm room if none of them overlap?

A) 90 B) 91 C) 117 D) 130

26.

27. If the ratio of each term to the term following it is constant, what is the next term in the sequence $\frac{1}{12}, \frac{1}{2}, 3$?

A) 6 B) 12 C) 18 D) 36

27.

28. At Burgertown I can buy 4 hot dogs and 5 burgers for \$60 or 2 hot dogs and 3 burgers for \$33. The cost of 1 burger is

A) \$6.00 B) \$6.50 C) \$7.00 D) \$7.50

28.

29. $99 \times 665 =$

A) $6 \times 9 \times 11 \times 65$ B) $100 \times 665 - 99$
C) $3 \times 7 \times 11 \times 15 \times 19$ D) $7 \times 11 \times 15 \times 19 \times 23$

29.

30. The difference between the sum of the positive odd integers less than 100 and the sum of the positive even integers less than 100 is

A) 49 B) 50 C) 99 D) 148

30.

31. Thirty seconds is what fraction of 1 hour?

A) $\frac{1}{2}$ B) $\frac{1}{60}$ C) $\frac{1}{120}$ D) $\frac{1}{240}$

31.

32. Robinson is stranded with an 80 ml vial of medicine. On the 1st day he sips 10 ml of it and replaces it with water, uniformly mixing it with the remaining medicine; on the 2nd day, he sips 15 ml of liquid from the vial. How much more medicine does he drink on the 2nd day than on the 1st?

A) $\frac{5}{2}$ ml B) $\frac{25}{8}$ ml C) $\frac{35}{8}$ ml D) 5 ml

32.

33. The perimeter of an isosceles triangle is 48. The altitude to its base divides it into two congruent triangles, each of perimeter 38. What is the length of the altitude to the triangle's base?

A) 14 B) 19 C) 24 D) 38

33.

34. What is the probability that a leap year has 53 Mondays?

A) 0 B) 1/7 C) 2/7 D) 3/7

34.

35. When fully expanded, how many digits does 20^{17} have?

A) 32 B) 28 C) 25 D) 23

35.

The end of the contest 7

Visit our Website at http://www.mathleague.com

Solutions on Page 73 • Answers on Page 138

Math League Press, P.O. Box 17, Tenafly, New Jersey 07670-0017

2017-2018 Annual 7th Grade Contest

Tuesday, February 20 (alternate date: February 27), 2018

7

Instructions

- **Time** Do *not* open this booklet until you are told by your teacher to begin. You might be *unable* to finish all 35 questions in the 30 minutes allowed.
- **Scores** Please remember that *this is a contest, and not a test*—there is no "passing" or "failing" score. Few students score as high as 28 points (80% correct). Students with half that, 14 points, *should be commended!*
- **Format, Point Value, & Eligibility** Every answer is an A, B, C, or D. Write answers in the *Answers* column. A correct answer is worth 1 point. Unanswered questions get no credit. You **may** use a calculator.

Answers

1. Pick any integer greater than 1. Double it twice, then triple the result. The final outcome is _?_ of your starting integer.

   A) 700% B) 1100% C) 1200% D) 1300%

   1.

2. Barry listened to the radio for 3 hours and 36 minutes. Rounded to the nearest 10 minutes, for how many minutes was Barry listening?

   A) 210 B) 220 C) 330 D) 340

   2.

3. Divide 99 by 22 to get a quotient and remainder. Divide that remainder by that quotient, and the new remainder is

   A) 4 B) 3 C) 2 D) 1

   3.

4. $10 - 8 \div 4 \times 6 + 2 =$ _?_

   A) 0 B) 2 C) 5 D) 12

   4.

5. The hundreds digit of the product of all whole numbers from 1 to 20 is

   A) 6 B) 4 C) 2 D) 0

   5.

6. Which of the following is the sum of the prime factors of 2018?

   A) 11 B) 219 C) 1011 D) 2019

   6.

7. If the length of the longest side of a triangle is 18, which of the following could *not* be the length of its second-longest side?

   A) 9 B) 10 C) 12 D) 17

   7.

8. My final score in a competition is the average of my scores on 5 rounds. To get a final score of 88 after getting 84, 80, and 92 on the first 3 rounds, what must be my average score for the last 2 rounds?

   A) 88 B) 90 C) 92 D) 96

   8.

9. How many even integers are between 19 and 91?

   A) 35 B) 36 C) 37 D) 71

   9.

10. Professor Peach teaches chemistry to clever kids. The ratio of freshmen to other students in his class is 3:8. The total number of students in Professor Peach's class could be

    A) 42 B) 45 C) 56 D) 77

    10.

11. $4^{40} \div 2^{20} =$

    A) 2^2 B) 2^4 C) 2^{20} D) 2^{60}

    11.

12. Two-dozen pens + 6 pens + 12 pens = _?_ pens

    A) 32 B) 36 C) 42 D) 48

    12.

13. I paid $5 and got 5 quarters, 5 dimes, and 5 nickels in change. I spent

    A) $3.00 B) $3.25 C) $3.45 D) $3.75

    13.

Go on to the next age ⟹ 7

Answers

14. One side of Todd's truck is a perfect rectangle with an area of 12 m^2. If its length is 3 times its width, then its perimeter is

A) 8 m B) 12 m C) 16 m D) 20 m

14.

15. If a bird in the hand is worth two in the bush, and a bird in the bush is worth four in the sky, then 4 birds in the hand are worth _?_ birds in the sky.

A) 1 B) 4 C) 16 D) 32

15.

16. On each of the four shelves of my bookcase is a different prime number of books. There could be a total of _?_ books on my shelves.

A) 15 B) 21 C) 22 D) 24

16.

17. Seven years ago I realized that my age would be tripled twelve years from then. How old am I now?

A) 11 B) 13 C) 16 D) 18

17.

18. How many fractions with a numerator of 1 and a whole-number denominator are greater than 0.01 and less than 1?

A) 98 B) 99 C) 100 D) 101

18.

19. If I write the letters R-E-P-E-A-T repeatedly, stopping when I have written exactly 100 letters, how many times do I write the letter E?

A) 16 B) 18 C) 32 D) 34

19.

20. On my map, 1 cm represents 100 km. If a park shown on the map is a rectangle that is 2.5 cm by 4 cm, the area of the actual park is _?_ km^2.

A) 100 B) 1000 C) 10000 D) 100000

20.

21. Gloomy Gus's Tuesday rain cloud shows up every Tuesday at 8:30 A.M. and every 50 minutes after that. Its last appearance on Tuesday is at _?_ P.M.

A) 11:00 B) 11:10 C) 11:30 D) 11:50

21.

22. If my lucky number divided by its reciprocal is 100, then the square of my lucky number is

A) 100 B) 10 C) 1 D) $\frac{1}{100}$

22.

23. Five consecutive even integers have a sum of 190. The largest of the integers is

A) 34 B) 38 C) 40 D) 42

23.

24. The ratio 7.8 to 5.2 is equivalent to

A) 8 to 5 B) 7 to 5 C) 3 to 2 D) 7 to 2

24.

25. 15% of 80 is 40% of

A) 30 B) 55 C) 105 D) 210

25.

Go on to the next page ⟹ 7

26. It took me 90 minutes to cycle 45 km to the beach. Later I got a ride from the beach to the park at twice my cycling speed. If the ride to the park took 15 minutes, what distance did I travel from the beach to the park?

A) 15 km B) 30 km C) 45 km D) 135 km

26.

27. Ted, Rick, and Sam painted a wall together. Ted painted 80% more of the wall than Sam painted. Sam painted 40% less than Rick. Ted painted __?__ of the amount that Rick painted.

A) 102% B) 108% C) 120% D) 140%

27.

28. The greatest integer power of 20 that is a divisor of 50^{50} is

A) 20^{20} B) 20^{25} C) 20^{50} D) 20^{125}

28.

29. The ones digit of the sum of all even integers from 2 to 1492 is

A) 2 B) 4 C) 8 D) 0

29.

30. The median of $\frac{1}{2}, \frac{1}{3}, \frac{1}{4}, \frac{1}{5}, \frac{1}{6}$, and $\frac{1}{7}$ is

A) $\frac{1}{8}$ B) $\frac{223}{840}$ C) $\frac{5}{14}$ D) $\frac{9}{40}$

30.

31. The average of all positive even integers from 2 to 2018 is

A) 1000 B) 1009 C) 1010 D) 1014

31.

32. Pirate Percy has 300 coins in his chest. Of the Spanish coins, 20% are gold. If 100 of the coins are gold but not Spanish and 70 of the coins are neither gold nor Spanish, how many Spanish gold coins are in Percy's chest?

A) 20 B) 26 C) 30 D) 34

32.

33. When I divided the population of my city by the number of streets in the city, I got a remainder of 18. If the exact quotient on my calculator was 123.06, how many streets are there in my city?

A) 60 B) 120 C) 186 D) 300

33.

34. What is the greatest number of 3-by-7 rectangles that can be placed inside an 80-by-90 rectangle with no overlapping?

A) 312 B) 330 C) 334 D) 342

34.

35. How many four-digit whole numbers have four different even digits and a ones digit greater than its thousands digit?

A) 36 B) 54 C) 60 D) 90

35.

The end of the contest 7

Visit our Website at http://www.mathleague.com

Solutions on Page 77 • Answers on Page 139

Math League Press, P.O. Box 17, Tenafly, New Jersey 07670-0017

2018-2019 Annual 7th Grade Contest

Tuesday, February 19 (alternate date: February 26), 2019

7

Instructions

- **Time** Do *not* open this booklet until told by your teacher to begin. You might be *unable* to finish all 35 questions in the 30 minutes allowed.
- **Scores** Please remember that *this is a contest, not a test*—there is no "passing" or "failing" score. Few students score as high as 28 points (80% correct). Students with half that, 14 points, *should be commended!*
- **Format, Point Value, & Eligibility** Every answer is an A, B, C, or D. Write answers in the *Answers* column. A correct answer is worth 1 point. Unanswered questions get no credit. You **may** use a calculator.

1. $(2 \times 4 \times 8) \div 2 = 4 \times$ _?_   1.

A) 2 B) 4 C) 8 D) 16

2. Al sleeps daily for 3 times as many hours as he is awake. For how many hours does Al sleep daily?   2.

A) 6 B) 9 C) 12 D) 18

3. The number 36 is the product of -6 and   3.

A) -6 B) 6 C) -30 D) 42

4. $20 \times 18 = 20 \times 19 + 20 \times$ _?_   4.

A) -1 B) 0 C) 1 D) 20

5. Angel arrived $\frac{3}{10}$ of an hour early for her noon appointment. At what time did Angel arrive?   5.

A) 11:18 a.m. B) 11:20 a.m. C) 11:40 a.m. D) 11:42 a.m.

6. The product of the least and greatest positive odd factors of 2019 is   6.

A) 673 B) 2019 C) 2020 D) 6057

7. The average value of the ten whole numbers from 0 through 9 is   7.

A) 5.5 B) 5 C) 4.5 D) 4

8. $2019 \times 3 + 2019 \div 3 = 2019 \times (3 +$ _?_$)$ 8.

A) 0 B) $\frac{1}{3}$ C) 1 D) 3

9. The product of four 4s equals the sum of _?_ 4s.   9.

A) 4 B) 3×4 C) 4^3 D) 4^4

10. What is the area of a square if one-third its side-length is 4?   10.

A) 12 B) 16 C) 48 D) 144

11. On a Monday my surf club had 20 members. If the number of members doubled each day, on what day did my club first have over 2018 members?   11.

A) Sunday B) Monday C) Tuesday D) Friday

12. Rounding a decimal to the nearest whole number yields a number that is at most _?_ greater than the original decimal.   12.

A) 0.05 B) 0.1 C) 0.5 D) 0.9

13. The perimeter of a rectangle with area 2019 and integral side-lengths is greatest when its length and width differ by   13.

A) 0 B) 1 C) 670 D) 2018

Go on to the next page ➡ **7**

14. If half of my pals have at least 1 pet, and 1/3 of my pals with a pet have more than 1 pet, what fraction of my pals have exactly 1 pet?

A) $\frac{1}{6}$ B) $\frac{1}{3}$ C) $\frac{2}{3}$ D) $\frac{5}{6}$ — 14.

15. The average of 0.5, 1.5, and 2.5 equals the average of 1 and

A) 1 B) 1.5 C) 2 D) 2.5 — 15.

16. $9 \times 90 \times 900 \times 9000 = 9 \times$ __?__

A) 100^3 B) 900^3 C) 9000^3 D) $9\,000\,000^3$ — 16.

17. What is one less than the product -18×19?

A) -341 B) -342 C) -343 D) -344 — 17.

18. When I divide the number of digits in the decimal form of 10^{2018} by 4, the remainder is

A) 3 B) 2 C) 1 D) 0 — 18.

19. My first name has 60% as many letters as my last name. My first name *could* be

A) Al B) Ali C) Alex D) Alexa — 19.

20. What is the *least* possible sum of two integers whose product is 12?

A) -13 B) -11 C) 7 D) 8 — 20.

21. Of the first 100 positive integers, __?__ are *not* multiples of both 2 and 3.

A) 16 B) 32 C) 64 D) 84 — 21.

22. If one-third of the eggs in each carton of 1-dozen eggs are cracked, I must buy __?__ cartons to get 16-dozen eggs that are *not* cracked.

A) 48 B) 36 C) 24 D) 20 — 22.

23. Which of the following is nearest in value to 8.25?

A) $8\frac{2}{5}$ B) $8\frac{2}{10}$ C) $8\frac{5}{10}$ D) $8\frac{10}{25}$ — 23.

24. I bowled on 2 days every week, on a different pair of days each week that I bowled. For at most how many weeks did I bowl?

A) 14 B) 21 C) 28 D) 35 — 24.

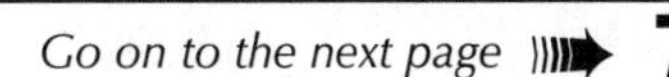

25. Which of the following has the least value?

A) 0.1 B) 0.01 C) 0.0011 D) $(0.01)^2$ — 25.

Go on to the next page ➡ **7**

26. A rectangular prism is 5 m long, 4 m wide, and 6 m high. What is the sum of the lengths of its edges?

    A) 15 m B) 60 m C) 80 m D) 120 m

26.

27. What is the ratio of $1\frac{1}{3}$ to its reciprocal?

    A) 1 **B)** $\frac{3}{4}$ C) $\frac{4}{3}$ D) $\frac{16}{9}$

27.

28. Pens come in packs of 3, 6, 8, and 12. I bought 12 packs and got a total of 121 pens. If I bought at least one of each size pack, how many packs of 8 pens did I buy?

    A) 1 B) 2 C) 3 D) 4

28.

29. $3^2 \times 8^2 \times 5^2 = 6^2 \times \underline{\ ?\ } \times 10^2$

    A) $\frac{1}{2}$ B) 2 C) 2^2 D) 2^3

29.

30. I wrote the first 100 positive integers in order, and then erased every "1" I had written. How many digits did I erase?

    A) 18 B) 19 C) 20 D) 21

30.

31. What is the difference between the product and the sum of the non-zero digits of 20^{10} when it is written in decimal form?

    A) 1 B) 2 C) 10^2 D) 2×10

31.

32. In the sequence 20, $\frac{19}{2}, \frac{18}{3}, \frac{17}{4}, \ldots$, each term after the first term is gotten by subtracting 1 from the previous term's numerator and adding 1 to the previous term's denominator. How many terms in this sequence are positive integers?

    A) 1 B) 2 C) 3 D) 4

32.

33. Two congruent rectangular cards partially overlap. The area of overlap is a square with area 4, and the total area of the regions of the faces of the two cards that *do not overlap* is 12. What is the area of one card?

    A) 4 B) 6 C) 8 D) 10

33.

34. If the mean of three positive integers is 5, then the product of all 3 integers is *at most*

    A) 105 B) 120 C) 125 D) 150

34.

35. What is the sum of the digits of the least 3-digit integer whose square is a 6-digit integer?

    A) 5 B) 7 C) 9 D) 11

35.

The end of the contest ✍ **7**

Visit our Website at http://www.mathleague.com

Solutions on Page 81 • Answers on Page 140

Math League Press, P.O. Box 17, Tenafly, New Jersey 07670-0017

2019-2020 Annual 7th Grade Contest

Tuesday, February 18 or February 25, 2020

7

Instructions

- **Time** Do *not* open this booklet until told by your teacher to begin. You might be *unable* to finish all 35 questions in the 30 minutes allowed.
- **Scores** Remember that *this is a contest, not a test*—there is no "passing" or "failing" score. Few students score 28 points (80% correct). Students with half that, 14 points, *should be commended!* High-scoring students may be invited to our "Math Camp" in July.
- **Format, Point Value, & Eligibility** Every answer is an A, B, C, or D. Write answers in the *Answers* column. A correct answer is worth 1 point. Unanswered questions get no credit. You **may** use a calculator. You're eligible for this contest only if you are in grade 7 or below and only if you don't also take this year's Annual 6th or Annual 8th Grade Contest.

1. $(2 \times 4 \times 8) \div 2 = 1 \times 2 \times 4 \times$ __?__

   A) 1 B) 2 C) 4 D) 8

2. Sam spent four-fifths of $80 on candy. Sam spent __?__ on candy.

   A) $16 B) $32 C) $64 D) $300

3. If the total value of my quarters equals the total value of 75 nickels, I must have exactly __?__ quarters.

   A) 3 B) 15 C) 25 D) 50

4. $100 \times 0.01 = 10 \times$ __?__

   A) 0 B) 0.01 C) 0.1 D) 1

5. Petunia has only cats, dogs, and birds as pets. If $\frac{1}{4}$ of her pets are cats and $\frac{1}{4}$ of the rest are dogs, what fraction of her pets are birds?

   A) $\frac{9}{16}$ B) $\frac{8}{16}$ C) $\frac{7}{16}$ D) $\frac{6}{16}$

6. My bib number rounded to the nearest 10 is 10 more than my bib number rounded to the nearest 100. My bib number could be

   A) 95 B) 100 C) 104 D) 105

7. The reciprocal of one-fourth equals __?__-fourths.

   A) four B) eight C) ten D) sixteen

8. Hanni's handstand lasted 480 seconds longer than Hal's 48-minute handstand. Hanni's handstand lasted __?__ minutes.

   A) 56 B) 58 C) 86 D) 96

9. If the product of 3 consecutive positive integers is a multiple of 10, the least possible value of their product is

   A) 10 B) 30 C) 60 D) 720

10. A square's perimeter is divisible by 12. Its area must be divisible by

    A) 6 B) 8 C) 9 D) 12

11. If 2019 is the greatest of 2020 consecutive integers, the product of these 2020 integers must be

    A) 0 B) negative C) prime D) odd

12. $20 \times 19 - 19 =$

    A) 0 B) 20 C) 19^2 D) 20^2

13. The sum of the digits of the greatest perfect square less than 1000 is

    A) 1 B) 4 C) 9 D) 16

Go on to the next page ⟹ 7

14. Carl counted to 2020^2 by 2s, starting with 2. How many numbers did he count?

A) 2020 B) 1010×1010 C) 1010×2020 D) 2020×2020

14.

15. The average number of fish per tank in my 4 fish tanks is 5. If each tank has at least 1 fish, what is the greatest possible number of fish in a tank?

A) 17 B) 18 C) 19 D) 20

15.

16. The least possible perimeter of a rectangle with integer side-lengths and area 144 is

A) 48 B) 50 C) 52 D) 60

16.

17. Ben bought 6 times as many \$6 books as \$8 books. If Ben bought 154 books, how much did he spend on \$8 books?

A) \$88 B) \$154 C) \$168 D) \$176

17.

18. If I divide 100 by the least integer greater than 1 that has a positive remainder when divided by 2, 3, 4, 5, 6, 7, 8, and 9, my remainder is

A) 1 B) 2 C) 3 D) 4

18.

19. Each year, I grow 10% of my greatest height the year before. At the end of 2022, I will be _?_ % taller than I will be at the end of 2020.

A) 11 B) 12 C) 20 D) 21

19.

20. If 6 is a factor of the product of 3 consecutive primes, what is the sum of the 3 primes?

A) 5 B) 6 C) 10 D) 30

20.

21. The expression 100^2 can be written as each of the following *except*

A) $(10^2)^2$ B) $(10 \times 10)^2$ C) $10^2 \times 10^2$ D) $(10^2 \times 10^2)^2$

21.

22. A list of the first 2020 positive integers contains exactly _?_ more multiples of 4 than multiples of 5.

A) 100 B) 101 C) 110 D) 111

22.

23. My faucet drips once every 7 minutes. If it dripped at noon yesterday, at what time did it first drip today?

A) 12:00 a.m. B) 12:01 a.m. C) 12:03 a.m. D) 12:06 a.m.

23.

24. The square of the square of a prime integer has _?_ positive factors.

A) 1 B) 3 C) 4 D) 5

24.

25. The number –0.1 is closest in value to

A) 0.1 B) $(0.1)^2$ C) $(0.1)^3$ D) 2×0.1

25.

Go on to the next page ⟫⟫➨ 7

26. Ky ran 400 m in half the time it took Cy to run 1 km. What is the ratio of Ky's speed to Cy's speed?
A) 2:5 B) 1:2 C) 4:5 D) 4:1 | 26.

27. A dozen people took turns picking a gumball from a jar of 2019 gumballs. They kept picking until not enough were left for each person to get an equal number of gumballs. There were __?__ gumballs left.
A) 1 B) 3 C) 7 D) 9 | 27.

28. I subtracted 1 from every prime less than 100. How many of the numbers I got are multiples of 5?
A) 5 B) 6 C) 7 D) 8 | 28.

29. Increasing the ones digit of each of my two addends by 1 would change both the tens digit and the hundreds digit of their current sum. The current sum of my addends could be
A) 189 B) 197 C) 198 D) 209 | 29.

30. -9 is greater than -99 by the same number that 9 is greater than
A) -81 B) -90 C) -99 D) -107 | 30.

31. Ali wrote the 100 smallest positive multiples of 2020 from least to greatest. What was the 240th digit that Ali wrote?
A) 0 B) 2 C) 4 D) 8 | 31.

32. Bai wrote the first 2020 positive integers on a blackboard, then erased all the multiples of 2, then all the multiples of 5. How many integers were still there?
A) 606 B) 808 C) 1212 D) 1414 | 32.

33. The least possible sum of 7 positive integers whose median is 4 is
A) 16 B) 19 C) 23 D) 28 | 33.

34. The expression 2^4 can be written as the product of identical powers of 2 in two ways: $2^1 \times 2^1 \times 2^1 \times 2^1$ and $2^2 \times 2^2$. In how many ways can 2^{2020} be written as the product of identical powers of 2?
A) 3 B) 6 C) 11 D) 1010 | 34.

35. A Ferris wheel has 24 cars spaced evenly apart and numbered clockwise from 1 to 24. The bottom car is #5. Which 3 cars could become the bottom car after the Ferris wheel turns at least $\frac{2}{3}$ and at most $\frac{3}{4}$ of a full clockwise rotation?
A) 11, 12, 13 B) 13, 14, 15 C) 17, 18, 19 D) 21, 22, 23 | 35.

The end of the contest ✍ **7**

Visit our Website at http://www.mathleague.com
Solutions on Page 85 • Answers on Page 141

Math League Press, P.O. Box 17, Tenafly, New Jersey 07670-0017

2020-2021 Annual 7th Grade Contest

Tuesday, February 16 or February 23), 2021

7

Instructions

- **Time** Do *not* open this booklet until you are told by your teacher to begin. You might be *unable* to finish all 35 questions in the 30 minutes allowed.
- **Scores** Please remember that *this is a contest, and not a test*—there is no "passing" or "failing" score. Few students score as high as 28 points (80% correct). Students with half that, 14 points, *should be commended!*

1. 2000 + 200 + 20 + 2 − __?__ = 2021.

   A) 1 B) 11 C) 21 D) 201

2. A team has bags with 3 bats and 9 baseballs in each. If there are 18 more baseballs than bats in the bags, the total number of balls is

   A) 27 B) 36 C) 48 D) 54

3. Which of the following is **not** equal to 0.5?

   A) 0.50 B) $\frac{1}{2}$ C) $\frac{50}{10}$ D) 50%

4. 2020 m = 1 km + __?__ m.

   A) 1020 B) 2202 C) 2222 D) 2244

5. 1/2 ÷ 1/4 ÷ 1/8 = 1/16 ÷ 1/32 ÷ __?__.

   A) 1/2 B) 1/8 C) 1/32 D) 1/64

6. In each classroom in my school there are 6 groups of chairs, with 3 chairs per group. If there are 270 chairs in all, there are __?__ classrooms.

   A) 15 B) 30 C) 45 D) 90

7. How many numbers between 100 and 200 are multiples of 9?

   A) 9 B) 10 C) 11 D) 12

8. When rounded to the nearest tenth, 2021 ÷ 1000 =

   A) 2.0 B) 2.02 C) 20.2 D) 20.21

9. Which of the following has the fewest factors of 2?

   A) 2000 B) 2200 C) 2220 D) 2222

10. The base of an isosceles triangle with integral sides is 10. What could be the perimeter?

    A) 19 B) 20 C) 21 D) 22

11. On a 35-question test, Pat correctly answered 2/5 of the first 10 questions and all the remaining ones. Pat answered __?__ of the 35 questions correctly.

    A) 14 B) 25 C) 29 D) 31

12. Emma counted up from 0 by 1s every second for 6 minutes, and then down by 1s every second for 2 minutes. What number did she end on?

    A) 120 B) 240 C) 360 D) 480

13. Which of the following expressions is closest to 0?

    A) 0.2 − 0.1 B) 0.21 − 0.12 C) 0.56 − 0.55 D) 0.661 − 0.599

Go on to the next page ⟹ 7

14. Beginning at 5:00 AM, my sprinkler turns on for 20 minutes, then turns off for 5 minutes, turns on again for 20 minutes, and then turns off again for 5 minutes. If my sprinkler continues this same pattern all morning, the water is off at

A) 8:06 AM B) 9:06 AM C) 10:06 AM D) 11:06 AM

14.

15. I practice catching footballs for 72 minutes every day. For what fraction of each day do I practice?

A) 1/3 B) 5/144 C) 1/18 D) 1/20

15.

16. Which expression is least?

A) $5^{20} \div 6^{20}$ B) $5^{21} \div 6^{21}$ C) $5^{22} \div 6^{23}$ D) $5^{23} \div 6^{22}$

16.

17. If my watch is accurate on New Year's Day but it loses 1 second per day, what month will it be when the watch is 12 minutes behind?

A) January B) June C) July D) December

17.

18. The product of 5.125 and _?_ is a whole number.

A) 5555 B) 6666 C) 7777 D) 8888

18.

19. 2020 is divisible by _?_ different prime numbers.

A) 2 B) 3 C) 4 D) 5

19.

20. Sally Speedster read 80 pages of a book at 2 pages each minute, and then 20 pages at 1 page each minute. On average, how many of the 100 pages did Sally read each minute?

A) 3/2 B) 8/5 C) 5/3 D) 7/4

20.

21. The smallest whole number that is divisible by 7 and leaves a remainder of 1 when divided by 2 **and** by 5 is between

A) 0 and 25 B) 26 and 50 C) 51 and 75 D) 76 and 100

21.

22. 50% of 50% = one quarter of

A) $\frac{1}{4}$ B) $\frac{1}{2}$ C) 1 D) 25%

22.

23. Which of the following numbers has the most divisors?

A) 49 B) 64 C) 72 D) 81

23.

24. $\left(\frac{9}{4}\right)^2 + 3^2 =$

A) $\left(\frac{15}{4}\right)^2$ B) 4^2 C) 5^2 D) $\left(\frac{21}{4}\right)^2$

24.

25. On a Sunday, 1547 beach balls were distributed among all the area beaches. If each beach had the same number of balls, there could be

A) 21 beaches B) 35 beaches C) 57 beaches D) 91 beaches

25.

Go on to the next page

26. I have exactly 15 coins, each of which is a nickel, dime, or quarter. If I have exactly 2 dollars, I have at least __?__ quarters.

    A) 4 B) 5 C) 6 D) 7

    26.

27. In 50 minutes of dance time, Mo and Jo danced together for only 30 seconds! Mo and Jo danced together for __?__ percent of the dance time.

    A) 1 B) 5 C) 6 D) 10

    27.

28. The 2021st digit after the decimal in the decimal expansion of $1 \div 7$ is

    A) 1 B) 2 C) 4 D) 5

    28.

29. I bowled four games and averaged 130 points per game. If I ignore my lowest game, my average was 140. If no two games had the same score, what is the highest single game score I could have bowled?

    A) 150 B) 169 C) 197 D) 217

    29.

30. The sum of the reciprocals of 3 and 4 is

    A) 1/7 B) 1/5 C) 7/12 D) 12/5

    30.

31. A rhombus has an angle that has twice the measure of an adjacent angle. What is the measure of the smaller angle?

    A) 45° B) 60° C) 72° D) 120°

    31.

32. Zoe starts on the top rung of a 16-rung ladder and Dafne starts on the bottom rung. Every minute, Zoe climbs down 5 rungs and then climbs up 4 rungs while Dafne climbs up 3 rungs and then down 2 rungs. During which minute will they first touch the same rung?

    A) 2nd B) 5th C) 7th D) 8th

    32.

33. The larger angle between the minute hand and the hour hand of a circular clock at 1:10 PM measures

    A) 300° B) 315° C) 330° D) 335°

    33.

34. How many prime numbers less than 100 have reciprocals that are terminating decimals?

    A) 1 B) 2 C) 3 D) 4

    34.

35. When 2^{80} is written as a decimal, the sum of its tens and ones digits is

    A) 9 B) 11 C) 13 D) 15

    35.

The end of the contest 7

Visit our Website at http://www.mathleague.com

Solutions on Page 89 • Answers on Page 142

8th Grade Contests

•••••••••••••••••••••••••

2016-2017 through 2020-2021

Math League Press, P.O. Box 17, Tenafly, New Jersey 07670-0017

2016-2017 Annual 8th Grade Contest

Tuesday, February 21 or 28, 2017

8

Instructions

- **Time** Do *not* open this booklet until you are told by your teacher to begin. You might be *unable* to finish all 35 questions in the 30 minutes allowed.
- **Scores** Please remember that *this is a contest, and not a test*—there is no "passing" or "failing" score. Few students score as high as 28 points (80% correct). Students with half that, 14 points, *should be commended!*
- **Format, Point Value, & Eligibility** Every answer is an A, B, C, or D. Write answers in the *Answers* column. A correct answer is worth 1 point. Unanswered questions get no credit. You **may** use a calculator.

1. $(2 \times 22 \times 222) \div (1 \times 11 \times 111) =$

   A) 2 B) $2+2+2$ C) $2 \times 2 \times 2$ D) $1+11+111$

2. $4 + 4 \div 4 \times 4 - 4 =$

   A) 0 B) 4 C) 8 D) 16

3. The safe deposit boxes at my bank have been numbered consecutively from 1 to 500. What percent of the box numbers are multiples of 50?

   A) 0.1% B) 0.2% C) 2% D) 10%

4. One-tenth of 1 = one-tenth more than __?__

   A) -0.1 B) 0 C) 0.1 D) 1

5. The reciprocal of eleven-tenths is closest in value to

   A) 0.11 B) 0.1 C) 1.1 D) 1.0

6. How many 3-digit integers can be written as the product of three consecutive primes?

   A) 0 B) 1 C) 2 D) 3

7. The sum of the digits of the least common multiple of 16 and 24 is

   A) 5 B) 8 C) 12 D) 16

8. How many factors of 2017^{2017} are odd?

   A) 0 B) 1 C) 2017 D) 2018

9. The ratio of the side-length of a square with area 64 cm^2 to the side-length of a cube with volume 64 cm^3 is

   A) 2:3 B) 1:1 C) 2:1 D) 4:1

10. If the numerical degree measure of each angle of a polygon is a multiple of 5 but not of 10, then the polygon could be a

    A) rhombus B) square C) triangle D) pentagon

11. When written as a fraction in lowest terms, what is the denominator of $\frac{1}{10} \div \left(\frac{1}{100} + \frac{1}{1000}\right)$?

    A) 10 B) 11 C) 1000 D) 1100

12. If the product of 2017 consecutive positive integers equals the product of 2016 consecutive positive integers, then the sum of the greatest integer from each group must be

    A) 4032 B) 4033 C) 4034 D) 4035

Answers
1.
2.
3.
4.
5.
6.
7.
8.
9.
10.
11.
12.

13. Each edge of a cube is parallel to exactly _?_ other edges of the cube. — 13.

A) 1 B) 2 C) 3 D) 4

14. The greatest possible integer length of a diagonal of a rectangle with area 12 and integer sides is — 14.

A) 5 B) 8 C) 12 D) 13

15. If 30% of them are quarters, the value of the coins in Al's bag of dimes and quarters could be — 15.

A) $10.25 B) $11.80 C) $13.05 D) $14.75

16. Which of the following is closest in value to 1? — 16.

A) 1.1 B) 1.1^{10} C) 1.1% D) 111%

17. Stan arranges chairs in rows, with the number of rows equal to the number of chairs in each row. If he has 999 chairs, what is the least number of chairs Stan must leave unused to form this arrangement? — 17.

A) 31 B) 33 C) 36 D) 38

18. How many factors of the product of any three primes are *not* prime? — 18.

A) 2 B) 3 C) 4 D) 5

19. Ed typed the first 10 000 positive integers, from least to greatest, in groups of 8. The sum of the integers in the 500th group he typed was — 19.

A) 30 864 B) 31 972 C) 32 144 D) 34 032

20. Tripling the length of each side of a cube multiplies the volume by — 20.

A) 3 B) 2^3 C) 3^2 D) 3^3

21. How many factors of 51^6 are prime? — 21.

A) 0 B) 1 C) 2 D) 3

22. In a group of 10 students, each student must shake hands with each of the 9 other students exactly once. There will be _?_ handshakes. — 22.

A) 19 B) 20 C) 45 D) 90

23. If one side of a square is a diameter of a circle with radius 2, the area of that part of the square outside the circle is nearest in value to — 23.

A) 2.28 B) 3.34 C) 3.43 D) 9.72

24. After Zak scored 100 on his sixth test, his average test score was 90. The average of his first 5 test scores was — 24.

A) 80 B) 85 C) 88 D) 92

25. The reciprocal of a number can never be — 25.

A) 0 B) 0.001 C) 1 D) 1 000 000

Go on to the next page ⟹ 8

Answers

26. What is the probability that a randomly chosen factor of 40 is odd?

A) $\frac{1}{5}$ B) $\frac{1}{4}$ C) $\frac{1}{3}$ D) $\frac{1}{2}$

27. Teenaged Ted's birthday is today. The sum of his age in months and his age in years is a perfect square. In how many years will the sum of Ted's age in months and his age in years next be a perfect square?

A) 3 B) 13 C) 26 D) 39

28. The chance of rain each day after Tuesday in my town is 50% of the chance of rain the day before. If the chance of rain on Tuesday is 100%, on what day does the chance of rain first decrease to less than 1%?

A) Sunday B) Monday C) Tuesday D) Wednesday

29. How many multiples of 2 between 1000 and 10000 contain only the digits 1, 2, 3, 4, and 5? (A digit may be used more than once or not used in a number.)

A) 64 B) 125 C) 128 D) 250

30. The ratio of the number of nuts to the number of raisins in my snack was 5:3. I ate 50 nuts but no raisins, and now the ratio is 3:2. I have _?_ more nuts than raisins.

A) 120 B) 150 C) 160 D) 180

31. Aya poured 1 m^3 of water into an empty rectangular-box-shaped aquarium with a 4 m × 4 m base and a height of 8 m. How deep was the water in the aquarium?

A) $\frac{1}{8}$ m B) $\frac{1}{4}$ m C) $\frac{1}{4\times4}$ m D) $\frac{1}{4\times4\times8}$ m

32. When written as a decimal, what is the sum of the digits of $2^{2017} \times 5^{2007}$?

A) 7 B) 8 C) 9 D) 10

33. My 4-letter password must contain 4 different letters. Of all possible passwords, what fraction contain the letters M, A, T, H, in any order?

A) $\frac{1}{23\times24\times25}$ B) $\frac{1}{23\times25\times26}$ C) $\frac{1}{23\times24\times26}$ D) $\frac{1}{24\times25\times26}$

34. Abe and George are the only candidates in an election. When 45% of voters mailed in votes, Abe got 20% fewer votes than George did. Abe must get _?_ of the remaining votes for the election to end in a tie.

A) $\frac{6}{11}$ B) $\frac{3}{5}$ C) $\frac{2}{3}$ D) $\frac{3}{4}$

35. I wrote the first 2017 positive integers in order from least to greatest. What was the 2017th digit I wrote?

A) 4 B) 5 C) 6 D) 7

The end of the contest 8

Visit our Website at http://www.mathleague.com

Solutions on Page 95 • Answers on Page 143

EIGHTH GRADE MATHEMATICS CONTEST

Math League Press, P.O. Box 17, Tenafly, New Jersey 07670-0017

2017-2018 Annual 8th Grade Contest

Tuesday, February 20 or 27, 2018

8

Instructions

- **Time** Do *not* open this booklet until you are told by your teacher to begin. You might be *unable* to finish all 35 questions in the 30 minutes allowed.
- **Scores** Please remember that *this is a contest, and not a test*—there is no "passing" or "failing" score. Few students score as high as 28 points (80% correct). Students with half that, 14 points, *should be commended!*
- **Format, Point Value, & Eligibility** Every answer is an A, B, C, or D. Write answers in the *Answers* column. A correct answer is worth 1 point. Unanswered questions get no credit. You **may** use a calculator.

1. $5 + 5 - 5 + 5 \times 5 + 5 \div 5 + 5 =$ | 1.

A) 0 B) 1 C) 25 D) 36

2. The product of 8 and 12 and their greatest common factor is | 2.

A) 24 B) 384 C) 768 D) 1804

3. How many positive integers less than 1000 are multiples of both 25 and 30? | 3.

A) 1 B) 2 C) 3 D) 6

4. Two players start 50 m apart and run toward each other at the same time in a straight line. If one runs 2m/sec. and the other runs 4 m/sec., in _?_ seconds they will be 32 m apart. | 4.

A) 2 B) 3 C) 4 D) 8

5. Drawing the diagonals of a rectangle creates exactly _?_ triangles. | 5.

A) 2 B) 4 C) 6 D) 8

6. The least possible average of 2017 different positive integers is | 6.

A) 1008 B) 1009 C) 2017 D) 2018

7. What is the sum of the two prime factors of 72? | 7.

A) 2 B) 3 C) 5 D) 6

8. Increasing a number by 20% is the same as multiplying it by | 8.

A) 20% B) 80% C) 120% D) 200%

9. $100 in nickels is _?_ more coins than $100 in dimes. | 9.

A) 100 B) 200 C) 1000 D) 2000

10. What is the range of any 2018 consecutive integers? | 10.

A) 1009 B) 2017 C) 2018 D) 2019

11. Written as a decimal, $\frac{123456789}{100}$ has exactly _?_ non-zero digits to the right of the decimal point. | 11.

A) 2 B) 3 C) 6 D) 7

12. Each choir member sang 1 song alone and 2 songs with the entire choir. If 24 songs were sung in all, the choir must have _?_ members. | 12.

A) 8 B) 11 C) 12 D) 22

13. A multiple of 2017 is divided by a multiple of 2018. What is the least remainder possible? | 13.

A) 0 B) 1 C) 2017 D) 2018

Go on to the next page ⟹ 8

14. My armful of identical gumballs weighs 4% less since I dropped one gumball. How many gumballs are in my arms now?

A) 23 B) 24 C) 25 D) 26

14.

15. The digits of the least 2-digit integer that is a perfect square *and* a perfect cube have the sum

A) 7 B) 8 C) 9 D) 10

15.

16. If two angles of a triangle are complementary, the triangle must be

A) equilateral B) right C) scalene D) obtuse

16.

17. Of the following, which has the greatest value?

A) $1 + 0.1$ B) $1^{10} + 0.1^{10}$ C) $1^{100} + 0.1^{100}$ D) $1^{1000} + 0.1^{1000}$

17.

18. The sum of the lengths of all the edges of a cube is 144 cm. What is the area of one face of the cube?

A) 144 cm^2 B) 196 cm^2 C) 256 cm^2 D) 324 cm^2

18.

19. The time 815 minutes after 8:15 P.M. is

A) 3:15 A.M. B) 9:50 A.M. C) 3:15 P.M. D) 9:50 P.M.

19.

20. The number 180 has _?_ more divisors than the number 120 has.

A) 0 B) 2 C) 30 D) 60

20.

21. The 8 houses on my street have consecutive integer addresses that add up to 1500. The address with the greatest numerical value is

A) 184 B) 187 C) 188 D) 191

21.

22. Which of these fractions is the sum of an integer and its reciprocal?

A) $\frac{7}{3}$ B) $\frac{8}{3}$ C) $\frac{9}{3}$ D) $\frac{10}{3}$

22.

23. The mixed number $2\frac{1}{4}$ is equivalent to many improper fractions that have integer numerators and denominators. The numerator of such a fraction could be any of the following except

A) 24 B) 27 C) 36 D) 45

23.

24. At my store, \$1 of every \$5 in sales is profit. If I split 10% of all profits equally among 10 people, each gets _?_% of the total sales.

A) 0.2 B) 2 C) 5 D) 20

24.

25. $2^{24} = \underline{\ ?\ } \times 2^{12}$

A) 2 B) 2^2 C) 2^{12} D) 2^{36}

25.

Go on to the next page ⟹ 8

26. Of the following, which expression has the least value? — 26.

A) $\frac{3^{100}}{4}$ B) $\left(\frac{3}{4}\right)^{100}$ C) $\frac{3}{4}$ D) $\frac{3}{4^{100}}$

27. I randomly select a positive integer less than 100. The probability that it is the product of exactly 3 different primes is — 27.

A) $\frac{1}{99}$ B) $\frac{4}{99}$ C) $\frac{5}{99}$ D) $\frac{8}{99}$

28. If the average of 3 consecutive ticket numbers is odd, then the sum of the least and greatest ticket numbers could be — 28.

A) 18 B) 20 C) 24 D) 28

29. Eve counted to 4^{60} by consecutive powers of 2, starting with $2^1, 2^2, 2^3, \ldots$. How many powers of 2 did Eve count? — 29.

A) 30 B) 120 C) 240 D) 3600

30. How many even integers between 1 and 1 000 000 have digits that are all primes? — 30.

A) 1365 B) 3906 C) 5400 D) 19 530

31. If 6 identical machines can fill 80 bottles of soda in 12 seconds, how many seconds would it take 36 of the same machines to fill 240 bottles of soda? — 31.

A) 6 B) 12 C) 18 D) 24

32. Of my 100 favorite released songs , 42% were released after the year 2015 and 76% were released before the year 2017. What percent of my favorite songs were released in 2016? — 32.

A) 18% B) 24% C) 34% D) 58%

33. (The number of positive even integers less than 10^6 that are perfect squares) : (the number of positive odd integers less than 10^6 that are perfect squares) = — 33.

A) 1:1 B) 2:1 C) 499:500 D) 999:1000

34. Of the following, which is a multiple of 4? — 34.

A) $2017^{2018}+1$ B) $2017^{2018}+3$ C) $2017^{2018}+5$ D) $2018^{2017}+1$

35. If the sum of the measures of two angles of a parallelogram is 108 degrees, the sum of the measures of three of its angles could be — 35.

A) 72 degrees B) 162 degrees C) 234 degrees D) 252 degrees

The end of the contest 8

Visit our Website at http://www.mathleague.com

Solutions on Page 99 • Answers on Page 144

Math League Press, P.O. Box 17, Tenafly, New Jersey 07670-0017

2018-2019 Annual 8th Grade Contest

Tuesday, February 19 or 26, 2019

8

Instructions

- **Time** Do *not* open this booklet until told by your teacher to begin. You might be *unable* to finish all 35 questions in the 30 minutes allowed.
- **Scores** Remember that *this is a contest, and not a test*—there is no "passing" or "failing" score. Few students score as high as 28 points (80% correct). Students with half that, 14 points, *should be commended!*

Format, Point Value, & Eligibility Every answer is an A, B, C, or D. Write answers in the *Answers* column. A correct answer is worth 1 point. Unanswered questions get no credit. You **may** use a calculator.

Answers

1. $(4 \times 6 \times 8 \times 10) \div (6 \times 8 \times 10) =$

   A) 3 B) 4 C) 12 D) $3 \times 6 \times 8 \times 10$

   1.

2. $(2 \div 3)$ rounded to the nearest hundredth is

   A) 0.33 B) 0.66 C) 0.67 D) 0.70

   2.

3. Baby Amy is one day older than Baby Barry. The product of their ages measured in days could be

   A) 33 B) 132 C) 245 D) 246

   3.

4. (The largest even divisor of 200) $\div$ (the largest odd divisor of 200) =

   A) 4 B) 8 C) 20 D) 200

   4.

5. An equilateral triangle with integer side-lengths has a perimeter that is numerically equal to the area of a square. Which of the following could be the length of a side of the square?

   A) 12 B) 10 C) 8 D) 4

   5.

6. I have only nickels, dimes, and quarters to pay for my dinner, which costs $12.60. The smallest number of coins I can use to pay is

   A) 51 B) 52 C) 54 D) 55

   6.

7. The smallest prime factor of 2019 is

   A) 1 B) 3 C) 19 D) 673

   7.

8. The product of four consecutive integers must be divisible by each of the following <u>except</u>

   A) 4 B) 6 C) 10 D) 12

   8.

9. There are __?__ hours in 4 weeks.

   A) 48 B) 96 C) 336 D) 672

   9.

10. If I divide my favorite number by its reciprocal, the quotient is 10 times as large as my favorite number. My favorite number is

    A) $\frac{1}{10}$ B) $\frac{1}{5}$ C) $\frac{1}{2}$ D) 10

    10.

11. The height of the smoke from my barbecue is 100 000 cm, which is the same as __?__ km.

    A) 1 B) 10 C) 100 D) 1000

    11.

12. If the degree measures of the angles of a triangle are in a 4:5:6 ratio, what is the difference between the measures of the largest and the smallest angles?

    A) 12° B) 24° C) 30° D) 36°

    12.

Go on to the next page ⟫⟫➔ 8

13. The population of a town started at 1000, then went up 10%, then down 20%, then back up 10%. The population of the town ended at

    A) 968 B) 972 C) 1000 D) 1024

    13.

14. In my orchard, there are 60 more apples than oranges, and 5 times as many apples as oranges. How many apples are there?

    A) 50 B) 75 C) 100 D) 125

    14.

15. A polygon in which every pair of angles is supplementary must be a

    A) triangle B) square C) rectangle D) hexagon

    15.

16. Which of the following is smallest in value?

    A) 2^{600} B) 3^{500} C) 4^{400} D) 5^{300}

    16.

17. $(2^{100} \times 4^{50}) \div 2 =$

    A) 2^{75} B) 2^{100} C) 2^{149} D) 2^{199}

    17.

18. What is the remainder when 3^{333} is divided by 10?

    A) 1 B) 3 C) 7 D) 9

    18.

19. On a series of tests, Gus got 100 once, 90 twice, and 80 five times. What was his average score for all of the tests?

    A) 80 B) 85 C) 90 D) 92

    19.

20. The product of the **thousands** and **tenths** digits of 1234.5678 is

    A) 5 B) 10 C) 35 D) 40

    20.

21. The probability of heads then tails then heads on 3 tosses of a coin is

    A) 0.125 B) 0.25 C) 0.375 D) 0.5

    21.

22. On January 1 last year, Rui got a jar of jellybeans. On each day he ate the same number of jellybeans. He counted 560 on January 31 before eating any and he counted 380 on March 17 before eating any. There were __?__ jellybeans in the jar when Rui got it.

    A) 600 B) 650 C) 680 D) 740

    22.

23. Jake used 120 boxes of tissues in 3 days! There are 144 tissues per box. That's __?__ tissues per minute!

    A) 2 B) 3 C) 4 D) 5

    23.

24. The number 5184 has __?__ positive odd divisors.

    A) 1 B) 2 C) 4 D) 5

    24.

25. The sum of 5 consecutive even integers could be

    A) 120 B) 125 C) 164 D) 212

    25.

Go on to the next page ➠ **8**

26. Jacques, who paints only smiley faces, signs and numbers each of his paintings. If he started with Smiley #1 and has painted through Smiley #111, how many times has he used the digit 1 in his numbering?

A) 12 B) 22 C) 24 D) 36

26.

27. How many whole numbers have squares that are between 2 and 200?

A) 12 B) 13 C) 24 D) 26

27.

28. A baker cuts circular cookies out of a flat rectangle of cookie dough. If the rectangle is 2 m by 1 m, and the cookies have radius 10 cm, at most how many cookies can the baker cut from the sheet of dough?

A) 50 B) 63 C) 64 D) 200

28.

29. 0.02% of 20% of _?_ = 200% of 2000

A) 1000 B) 100 000 C) 1 000 000 D) 100 000 000

29.

30. A miner combines 1200 kg of ore that is on average 3% gold with 2400 kg of ore that is on average 6% gold. If the 100 kg containing the most gold of the 3600 kg is 40% gold, the remaining ore will be _?_ gold.

A) 2% B) 3% C) 4% D) 5%

30.

31. Including face diagonals, the total number of diagonals of a cube is

A) 12 B) 14 C) 16 D) 24

31.

32. How many odd 3-digit integers greater than 500 are composed of 3 different non-zero digits?

A) 154 B) 175 C) 185 D) 200

32.

33. If I square all whole-number factors of 36 and multiply the resulting numbers, the product will be equal to

A) 36^2 B) 36^4 C) 36^8 D) 36^9

33.

34. When the four members of the Beaverton family carry a log, each has a 0.02 probability of tripping, and each probability is independent of the others. What is the probability that they will carry the log without any of them tripping?

A) $1-(0.02)^4$ B) $(0.98)^4$ C) $(0.02)^4$ D) $1-(0.98)^4$

34.

35. What is the largest prime factor of the product of all even numbers from 2 through 200?

A) 47 B) 97 C) 199 D) 2019

35.

The end of the contest 8

Visit our Website at http://www.mathleague.com

Solutions on Page 103 • Answers on Page 145

Math League Press, P.O. Box 17, Tenafly, New Jersey 07670-0017

2019-2020 Annual 8th Grade Contest

Tuesday, February 18 or 25, 2020

8

Instructions

- **Time** Do *not* open this booklet until you are told by your teacher to begin. You might be *unable* to finish all 35 questions in the 30 minutes allowed.
- **Scores** Please remember that *this is a contest, and not a test*—there is no "passing" or "failing" score. Few students score as high as 28 points (80% correct). Students with half that, 14 points, *should be commended!*
- **Format, Point Value, & Eligibility** Every answer is an A, B, C, or D. Write answers in the *Answers* column. A correct answer is worth 1 point. Unanswered questions get no credit. You **may** use a calculator.

2019-2020 8TH GRADE CONTEST

Answers

1. $8\,000\,000 \times 16\,000\,000 = \underline{\ ?\ } \times 1\,000\,000\,000\,000.$

   A) 24 B) 32 C) 64 D) 128

   1.

2. Triplet sisters are celebrating their birthday today! Which of the following could be the sum of their three ages?

   A) 6 B) 13 C) 17 D) 80

   2.

3. The number of fish in a school is equal to the cube of an integer. There could be _?_ fish in the school.

   A) 16 B) 27 C) 36 D) 101

   3.

4. Lu is riding her bike at 10 m/sec. What is her speed in km/hr.?

   A) 10 B) 36 C) 60 D) 72

   4.

5. 0.3% of 30% of 30 000 is

   A) 27 B) 270 C) 2700 D) 27 000

   5.

6. The greatest prime factor of 2020 is

   A) 2 B) 5 C) 101 D) 202

   6.

7. The sum of all the integers from 1 through 2019 is _?_ less than the sum of all the integers from 2 through 2020.

   A) 1 B) 2 C) 2019 D) 2020

   7.

8. The reciprocal of 6 divided by the reciprocal of 3 is

   A) $\frac{1}{18}$ B) $\frac{1}{2}$ C) 2 D) 18

   8.

9. $(2^{2019} \times 2^{2020})^2 =$

   A) 2^{4041} B) 2^{6059} C) 2^{8078} D) $2^{4078382}$

   9.

10. The smallest positive integer that can be divided by 36 and 75 with remainders of 0 has a square root of

    A) 30 B) 50 C) 52 D) 2700

    10.

11. The ratio of whole to bitten apples in a bowl is 5:2. If there are 9 more whole apples than bitten apples, how many apples are bitten?

    A) 3 B) 6 C) 15 D) 18

    11.

12. A rectangle with a perimeter of 24 has a maximum area of

    A) 20 B) 24 C) 36 D) 144

    12.

13. What is the date exactly 10 000 minutes before a new calendar year?

    A) Dec. 27 B) Dec. 26 C) Dec. 25 D) Dec. 24

    13.

Go on to the next page ⟹ **8**

14. Seville the barber has 20 human clients averaging 80 kg each, and 30 porcupine clients averaging 10 kg each. What is the average weight of one of these 50 clients?

    A) 38 kg B) 42 kg C) 45 kg D) 52 kg

15. $-2^4 + 2^4 \times (-2)^4 + 2^4 = $ _?_.

    A) 16 B) 256 C) 288 D) 528

16. The remainder when $(3^{33} + 5^{55} + 7^{77})$ is divided by 2 is

    A) 0 B) 1 C) 2 D) 3

17. The difference between the hundredths and the thousands digits of 8765.4321 is

    A) 1 B) 2 C) 4 D) 5

18. I bought shoes at a local store for \$60, 20% more than the online price. If online orders also have a shipping fee of 10% of the online price, I paid _?_ more than the total of the online price and the shipping fee.

    A) \$5.00 B) \$6.00 C) \$7.20 D) \$10.00

19. What is the ones digit of 19^{2020}?

    A) 9 B) 7 C) 3 D) 1

20. Tripling the length of each side of a rectangle multiplies the perimeter of the rectangle by

    A) 3 B) 6 C) 9 D) 12

21. Paola's age is a number that has exactly 5 positive divisors. Paola's age must be

    A) a prime squared B) a non-prime squared
    C) an odd number D) an even number

22. From 6 unique pies, I can pick _?_ different possible combos of 3 pies.

    A) 18 B) 20 C) 40 D) 120

23. The number on Bo's hat, the 2020th number in the sequence 8100, 8096, 8092, 8088,..., is

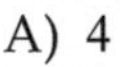

    A) 4 B) 20 C) 24 D) 64

24. In 2019, Min spent 200 hours playing video games. She played 250% more hours in 2018 than in 2019. Min played for _?_ hours in 2018.

    A) 450 B) 500 C) 600 D) 700

25. If $a \bigstar b = 2a + b^4$, then $8 \bigstar 2 =$

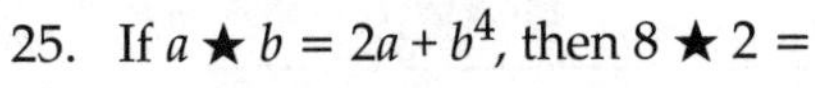

    A) 2 B) 2^4 C) 2^5 D) 2^8

Answers
14.
15.
16.
17.
18.
19.
20.
21.
22.
23.
24.
25.

Go on to the next page ⇛ 8

Answers

26. Each monster in a movie wears an integer, and the integers worn are consecutive. There are 6 monsters, so the sum of the integers could be

A) 67 B) 72 C) 75 D) 86

26.

27. How many integers between 100 and 1000 have exactly two 2s and no 3s as digits?

A) 23 B) 24 C) 26 D) 27

27.

28. At a zoo there are three times as many cats as crocodiles, and five times as many koalas as cats. The ratio of koalas to cobras is 9:4. What is the ratio of cobras to crocodiles?

A) 4:3 B) 20:3 C) 45:4 D) 135:4

28.

29. How many isosceles triangles with integer sides of lengths less than 21 have a side of length 13?

A) 14 B) 19 C) 32 D) 33

29.

30. Among 120 athletes at my school, the 3 most popular sports are baseball (played by 57 of them), basketball (played by 52), and soccer (played by 41). If 5 athletes play all three sports and 10 play none of them, how many athletes play exactly 2 of the 3 most popular sports?

A) 25 B) 30 C) 35 D) 40

30.

31. Each of my 10 friends thinks of any one of the 15 integers from -7 through 7, inclusive. Then they write down their integers and multiply them. What is the lowest possible product of their integers?

A) $(-7)^{10}$ B) $(-7)^{9}$ C) 0 D) -7^{10}

31.

32. Sailors Flo and Jet are at sea with enough food to last for 30 days together. If Flo were alone, the food would last 50 days. If Jet were alone, the food would last _?_ days.

A) 10 B) 40 C) 70 D) 75

32.

33. The remainder when 8^{9876} is divided by 5 is

A) 1 B) 2 C) 3 D) 4

33.

34. My dad has 54.36 times as many coins as I do. Dividing his number of coins by my (nonzero) number yields a minimum remainder of

A) 1 B) 3 C) 9 D) 36

34.

35. If 3 different integers are randomly chosen from the 20 smallest positive integers, what is the probability that their product is even?

A) 2/19 B) 1/8 C) 7/8 D) 17/19

35.

The end of the contest 8

Visit our Website at http://www.mathleague.com

Solutions on Page 107 • Answers on Page 146

Math League Press, P.O. Box 17, Tenafly, New Jersey 07670-0017

2020-2021 Annual 8th Grade Contest

Tuesday, February 16 or 23, 2021

8

Instructions

- **Time** Do *not* open this booklet until you are told by your teacher to begin. You might be *unable* to finish all 35 questions in the 30 minutes allowed.
- **Scores** Please remember that *this is a contest, and not a test*—there is no "passing" or "failing" score. Few students score as high as 28 points (80% correct). Students with half that, 14 points, *should be commended!*
- **Format, Point Value, & Eligibility** Every answer is an A, B, C, or D. Write answers in the *Answers* column. A correct answer is worth 1 point. Unanswered questions get no credit. You **may** use a calculator.

1. $222 \times 222 \times 222 = 111 \times 111 \times 111 \times$ _?_ — 1.

   A) 2   B) 2×3 C) 222   D) 2^3

2. What is the hundredths digit of the product 123456789×0.00001? — 2.

   A) 3   B) 4   C) 5   D) 6

3 A cup that holds 2 ℓ of tea can hold _?_ ml of tea. — 3.

   A) 20   B) 200   C) 2000   D) 20 000

4. The sum of the digits of the greatest 4-digit integer that contains exactly 3 different digits is — 4.

   A) 30   B) 31   C) 32   D) 33

5. How many positive integers less than 100 are not divisible by 3? — 5.

   A) 33   B) 34   C) 66   D) 67

6. Jan hiked half of a 36-km long trail in 3 hrs. Her average speed was — 6.

   A) 6 km/hr.   B) 9 km/hr.   C) 12 km/hr.   D) 18 km/hr.

7. How many multiples of 8 are factors of 8^2? — 7.

   A) 3   B) 4   C) 6   D) 8

8. The greatest 3-digit perfect square is _?_ more than the greatest 2-digit perfect square. — 8.

   A) 870   B) 880   C) 890   D) 900

9. 10% of 100 is 10 less than 10% of — 9.

   A) 20   B) 110   C) 120   D) 200

10. I wrote consecutive multiples of 2 in increasing order, starting with 2. What was the 10th digit I wrote? — 10.

   A) 1   B) 2   C) 3   D) 4

11. Today Al bought 36 hats and now has triple the number of hats he had yesterday. If tomorrow Al buys triple the number of hats he has today, how many hats will Al have after tomorrow's purchase? — 11.

   A) 108   B) 144   C) 162   D) 216

12. The sum of 5 integers is divisible by 20. The average must be divisible by — 12.

   A) 4   B) 5   C) 20   D) 25

13. Ed is late to work 5 days a month. If he works every day in April, the probability Ed gets to work on time on a given day in April is — 13.

   A) $\frac{1}{6}$ B) $\frac{1}{2}$ C) $\frac{2}{3}$ D) $\frac{5}{6}$

Go on to the next page ⟹ **8**

14. Eight friends split 6 identical pizzas evenly. If each friend got only whole slices, what was the lowest possible number of slices in each pizza?

A) 2 B) 3 C) 4 D) 6

14.

15. The reciprocal of 1.25 is

A) 0.125 B) 0.14 C) 0.4 D) 0.8

15.

16. If the difference between 2 prime integers is prime, what is the least possible sum of these integers?

A) 5 B) 7 C) 8 D) 9

16.

17. If the degree measures of the 3 angles of a triangle are consecutive integers, what is the sum of the measures of the 2 smallest angles?

A) 118° B) 119° C) 120° D) 121°

17.

18. May painted 6 pots in 2 days working alone. Elsa and May together painted 10 more pots over the next 3 days. Each paints at a constant rate. How many days would it take Elsa to paint 16 pots by herself?

A) 12 B) 24 C) 36 D) 48

18.

19. Exactly 3 diagonals can be drawn from each vertex of a regular _?_.

A) octagon B) parallelogram C) hexagon D) pentagon

19.

20. The difference between the squares of 2 consecutive integers is 25. What is the sum of these squares?

A) 221 B) 265 C) 313 D) 365

20.

21. How many different primes are factors of the product of the first 9 positive integers?

A) 4 B) 5 C) 9 D) 13

21.

22. A circle's diameter is what fraction of the circle's circumference?

A) $\frac{1}{\pi}$ B) $\frac{2}{\pi}$ C) $\frac{1}{2\pi}$ D) $\frac{\pi}{2}$

22.

23. How many of the 24 smallest integers greater than 26 are divisible by at least one of the 24 smallest integers greater than 1?

A) 17 B) 18 C) 19 D) 20

23.

24. If the length and width of a painter's canvas are in the ratio 3:2, the ratio of its perimeter to its width is

A) 3:2 B) 5:2 C) 5:1 D) 9:4

24.

25. What is the least possible value of a multiple of 6^3 that is also a multiple of 8^3?

A) 2×6^3 B) $2^3 \times 6^3$ C) $2 + 6^3$ D) $2^6 \times 6^3$

25.

Go on to the next page ⟹ 8

26. I have 25 nickels and 15 dimes to divide into 2 or more stacks of equal total value. What is the least number of stacks I can make if I use all 40 coins?

A) 3 B) 4 C) 5 D) 10

26.

27. For how many 2-digit positive integers less than 50 is it true that both digits are factors of the integer?

A) 9 B) 13 C) 17 D) 18

27.

28. The sum of 2^{2021} and 2^{2020} has the same value as

A) 2×2^{2020} B) 3×2^{2020} C) 4×2^{2020} D) $2^{2020} \times 2^{2020}$

28.

29. What is the ones digit of $9^9 + 9^{10} + 9^{11} + 9^{12} + 9^{13} + 9^{14} + 9^{15} + 9^{16} + 9^{17}$?

A) 9 B) 7 C) 1 D) 49

29.

30. Aida has 4 letter tiles: 2 As, 1 D, and 1 I. If Aida chooses the 4 tiles one at a time without looking, what is the probability she chooses them in the order A-I-D-A?

A) $\frac{1}{16}$ B) $\frac{1}{12}$ C) $\frac{1}{8}$ D) $\frac{1}{4}$

30.

31. Each of 200 runners ran nonstop for 45 seconds. After the 1st runner, each runner began to run when the previous one had run for 15 seconds. How much time elapsed between the time the 1st runner stopped running and the time the last runner stopped running?

A) 2985 sec. B) 5970 sec. C) 8940 sec. D) 8955 sec.

31.

32. For 3 years in a row on Dec. 31, interest of 10% on all the money in my account was added to the account. If I began on Jan. 1 with \$10000 and never withdrew, I earned _?_ interest total in 3 years.

A) \$3000 B) \$3100 C) \$3310 D) \$3641

32.

33. If 21^{100} has 101^2 positive divisors, how many does 25×21^{100} have?

A) 2×101^2 B) 3×101^2 C) 5×101^2 D) 25×101^2

33.

34. If the sum of the degree measures of the 2 smallest angles of a triangle is 102° and the sum of the measures of the 2 largest angles is 130°, what is the difference between the measures of the 2 largest angles?

A) 2° B) 26° C) 28° D) 32°

34.

35. For each pair of consecutive 2-digit positive integers, there is a remainder when you square the larger integer and divide it by the smaller integer. What is the sum of the remainders for all such pairs?

A) 44 B) 45 C) 89 D) 90

35.

The end of the contest **8**

Visit our Website at http://www.mathleague.com

Solutions on Page 111 • Answers on Page 147

Algebra Course 1 Contests

2016-2017 through 2020-2021

Math League Press, P.O. Box 17, Tenafly, New Jersey 07670-0017

2016-2017 Annual Algebra Course 1 Contest

Spring, 2017

Instructions

- **Time** Do *not* open this booklet until you are told by your teacher to begin. You will have only *30 minutes* working time for this contest. You might be *unable* to finish all 30 questions in the time allowed.
- **Scores** Please remember that *this is a contest, and not a test*—there is no "passing" or "failing" score. Few students score as high as 24 points (80% correct). Students with half that, 12 points, *should be commended!*
- **Format and Point Value** Every answer is an A, B, C, or D. Write answers in the *Answers* column. A correct answer is worth 1 point. Unanswered questions get no credit. You **may** use a calculator.

2016-2017 ALGEBRA COURSE 1 CONTEST

Answers

1. What are all values of x for which $\sqrt{x}$ is a real number?

   A) $-1 < x < 1$ B) $x \geq 0$ C) $x < 0$ D) $x > 0$

2. $(j + k)^2 = j^2 + k^2 + \underline{\ ?\ }$

   A) 0 B) jk C) $2jk$ D) $j + k$

3. The tuba band played a total of $t + u + b + a$ tunes. If t, u, b, and a are consecutive multiples of 2 and $t < u < b < a$, then the tuba band played exactly $\underline{\ ?\ }$ tunes.

   A) $t + 4$ B) $4t + 4$
   C) $t + 12$ D) $4t + 12$

4. If the value of xy is 24 when x is 3, what is the value of y^2?

   A) 72 B) 64 C) 16 D) 8

5. If 3 more than x equals 7, then $7 = \underline{\ ?\ }$.

   A) $x + 3$ B) $x - 3$ C) $3x$ D) $3 - x$

6. For any integer $x > 0$, the greatest common factor of $24x$ and $(2x+4x+6x)$ is

   A) $2x$ B) $4x$ C) $6x$ D) $12x$

7. For how many values of x does $\dfrac{1}{x^2 - 36}$ have no real values?

   A) 2 B) 3 C) 4 D) 6

8. If $x^3 = y^2$ for real numbers x and y, then x must be

   A) non-negative B) even C) a perfect square D) odd

9. What is the remainder when $x^2 - 9 + x^2$ is divided by $(x + 3)(x - 3)$?

   A) -9 B) 0 C) 9 D) x^2

10. The sum of my slowest and fastest scooter speeds is $(x + 5)^2$ km/hr. If my fastest speed is $(x^2 + 25)$ km/hr., what is my slowest speed?

    A) x km/hr. B) $5x$ km/hr.
    C) $10x$ km/hr. D) $25x$ km/hr.

11. $x \times x^2 \times x^3 \times x^4 = x^5 \times \underline{\ ?\ }$

    A) x^{21} B) x^5 C) x^4 D) 1

12. $(s - e) - (e - s) =$

    A) 0 B) $s - 2e$ C) $2s - 2e$ D) $2s - e$

Answers
1.
2.
3.
4.
5.
6.
7.
8.
9.
10.
11.
12.

Go on to the next page ➠ **A**

Answers

13. For all $x > 0$, each of the following is equivalent to 2^{2x} *except*

A) 2×2^x B) $(2 \times 2)^x$ C) $(2^2)^x$ D) $(2^x)^2$

13.

14. $|x^2 - 2x + 1| =$

A) $|1 - x|$ B) $|x - 1|$ C) $|(1 - x)^2|$ D) $x^2 + 2x + 1$

14.

15. If x and y are integers and $y < 2017$, how many positive values of x could satisfy $x^2 < y$?

A) 44 B) 45 C) 88 D) 89

15.

16. When I play darts, I score 5 points for every bullseye, plus an additional 3 points for each bullseye made after the 10th one. If b is the number of bullseyes I make after my 10th bullseye, my total score for bullseyes is

A) $8b$ B) $50 + 3b$
C) $50 + 8b$ D) $50 + 15b$

16.

17. What is the greatest possible sum of the numerical area and perimeter for a square with numerical perimeter 3 more than its numerical area?

A) 5 B) 9 C) 12 D) 21

17.

18. The product of the integral slopes of two parallel lines could be

A) 181 B) 183 C) 187 D) 289

18.

19. What is the least common denominator of $\frac{1}{x^2 - 1}$ and $\frac{1}{(x-1)^2}$?

A) $(x^2 - 1)(x + 1)$ B) $(x^2 - 1)(x - 1)^2$

C) $(x + 1)(x - 1)^2$ D) $(x + 1)(x - 1)^3$

19.

20. If f is a function such that $f(5) = 5$, then $f(f(5)) =$

A) 1 B) 5 C) 10 D) 25

20.

21. Anna rode exactly 1 km on her unicycle. If her tire made exactly 80 complete revolutions, what is the diameter of her tire?

A) $\frac{1}{80\pi}$ km B) $\frac{80}{\pi}$ km

C) $\frac{\pi}{80}$ km D) 80π km

21.

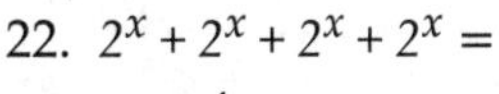

22. $2^x + 2^x + 2^x + 2^x =$

A) 2^{4x} B) 2^x C) 2^{2+x} D) 2^{2x}

22.

Go on to the next page ⟹ A

23. How many integers x, $0 \le x \le 10$, are there such that $\frac{(x-1)(x-2)(x-3)(x-4)}{(x-1)(x-2)(x-3)(x-4)} = 1$?

A) 7 B) 8 C) 9 D) 10

23.

24. Every second of my stop-motion movie "About a Bird" consists of 15 frames. How many frames do I need to make a 2-hour stop-motion movie?

A) 30 B) 1800 C) 72 000 D) 108 000

24.

25. If $n!$ represents the product of the first n positive integers, for what value of n does $\frac{n!}{(n-2)!} = 8010$?

A) 60 B) 70 C) 80 D) 90

25.

26. Before I ate, the ratio of noodles to veggies in my vegetable soup was 4:1. Now that I have eaten 50 noodles and 8 veggies, the new ratio is 5:2. What is the sum of the number of noodles and the number of veggies that are left in my vegetable soup?

A) 42 B) 58 C) 98 D) 100

26.

27. What is the product of all real solutions of $\frac{(x^2-25)(x^2-16)(x^2-9)(x^2-4)(x^2-1)}{(x+1)(x+2)(x+3)(x+4)(x+5)} = 0$?

A) 0 B) 5! C) -(5!) D) $5! \times 5!$

27.

28. Darth fills egg cartons at a constant rate of 24 cartons every quarter of an hour. Luke fills egg cartons 3 times as fast as Darth. Working together, how many egg cartons can Darth and Luke fill in 8 hours?

A) 576 B) 768 C) 2304 D) 3072

28.

29. I need to make up a secret 7-digit passcode. All digits must be different; no consecutive numbers can be next to each other; no even digit can be next to an even digit; and no odd digit can be next to an odd digit. What is the fourth digit of the greatest possible passcode that meets this criteria?

A) 6 B) 7 C) 8 D) 9

29.

30. For an integer n, $1 < n < 5$, at **most** how many factors of $((n^2)^2)^2$ are perfect squares?

A) 16 B) 9 C) 8 D) 4

30.

The end of the contest  **A**

Visit our Web site at http://www.mathleague.com

Solutions on Page 117 • Answers on Page 148

Math League Press, P.O. Box 17, Tenafly, New Jersey 07670-0017

2017-2018 Annual Algebra Course 1 Contest

Spring, 2018

A

Instructions

- **Time** Do *not* open this booklet until you are told by your teacher to begin. You will have only *30 minutes* working time for this contest. You might be *unable* to finish all 30 questions in the time allowed.
- **Scores** Please remember that *this is a contest, and not a test*—there is no "passing" or "failing" score. Few students score as high as 24 points (80% correct). Students with half that, 12 points, *should be commended!*
- **Format and Point Value** This is a multiple-choice contest. Each answer is an A, B, C, or D. Write each answer in the *Answer Column* to the right of each question. A correct answer is worth 1 point. Unanswered questions receive no credit. You **may** use a calculator.

Answers

1. If $a + l + g + e + b + r + a = 38$ and $a + b + a + g + e + l = 21$, then what is the value of r?

   A) 1 B) 10 C) 17 D) 59

   1. ____

2. Harald the Six Fingered traveled x km to raid the nearest kingdom. He traveled _?_ m.

   A) $100x$ B) $1000x$ C) $10\,000x$ D) $1\,000\,000x$

   2. ____

3. $(a + 2) + (2a + 4) - (3a - 6) - (4a + 8) =$

   A) $4a - 8$ B) $4a + 4$ C) $-4a - 8$ D) $-4a + 4$

   3. ____

4. Which of the following is NOT a factor of $b^4 - 81$?

   A) $b - 3$ B) $b + 3$ C) $b - 9$ D) $b^2 - 9$

   4. ____

5. If $s^{-11} > s^{-12}$, then s could equal

   A) 2 B) -2 C) $\frac{1}{2}$ D) $-\frac{1}{2}$

   5. ____

6. If $60s + 90t = 120s$, then $s - t =$

   A) $0.5t$ B) $2t$ C) $15t$ D) $30t$

   6. ____

7. $\frac{x^2}{y^2} \div \frac{x^1}{y^1} =$

   A) xy B) $\frac{x}{y}$ C) $\frac{y}{x}$ D) $\frac{x^3}{y^3}$

   7. ____

8. $(n - 4)^2 - (4 - n)^2 =$

   A) 0 B) $-16n$ C) $16n$ D) $2n^2 - 16n$

   8. ____

9. If p is a three-digit prime number, then which of the following could also be prime?

   A) $2p$ B) $3p$ C) $p - 2$ D) $p + 3$

   9. ____

10. My dog will catch up to the slowest member of the band in s seconds. If $s^2 + 64 = 16s$, then my dog will catch up in _?_ seconds.

    A) 2 B) 4 C) 6 D) 8

    10. ____

11. $(z^2 \times z^4 \times z^6 \times z^8)^2 = z^2 \times$ _?_

    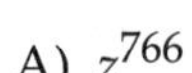

    A) z^{766} B) z^{40} C) z^{38} D) z^{20}

    11. ____

12. The line $3x - 4y = 5$ is parallel to the line

    A) $2x - 7y = 5$ B) $4y - 3x = 5$ C) $3x - 6y = 7$ D) $2x - 4y = 6$

    12. ____

Go on to the next page ➠ **A**

13. There are $4x$ boys and $5y$ girls in my class, and the ratio of boys to girls is 4:5. The ratio x:y is
A) 4:5 B) 5:4 C) 1:1 D) 16:25

13.

14. If the average of r, s, and t is 20 and the average of r and t is 24, then the value of s is
A) 4 B) 12 C) -4 D) -12

14.

15. Mrs. Robinson's son begs her for a cookie every $42m$ minutes, and her dog begs her for a cookie every $12m$ minutes, where m is a prime greater than 7. If they both beg for a cookie at noon, the next time they both beg for a cookie at the same time will be _?_ minutes later.
A) 30 B) $54m$ C) $84m$ D) $504m^2$

15.

16. If $a \Diamond b = 2a^2 + 3ab$, then $4 \Diamond 5 =$
A) 18 B) 24 C) 60 D) 92

16.

17. If Delaney drives at $18x$ km/hr., she drives at _?_ m/sec.
A) $5x$ B) $9x$ C) $18x$ D) $36x$

17.

18. If n is a positive integer, which of the following must be divisible by 6?
A) $(n+1)(n+3)(n+5)$ B) $(n+2)(n+4)(n+6)$
C) $(n+2)(n+5)(n+8)$ D) $(n+1)(n+3)(n+8)$

18.

19. What is the sum of the solutions to $|10-4x| = 5$?
A) 1.25 B) 3.75 C) 5 D) 10

19.

20. If $2^x \times 4^{2x} \times 8^{3x} = 2^y$, then $y =$
A) $2x^3$ B) $6x$ C) $6x^3$ D) $14x$

20.

21. Each time Alan falls asleep, he sleeps for exactly $8m$ minutes and then is awake for the next $4m$ minutes. If he falls asleep for the 1st time at 11 P.M. and wakes from his 6th time asleep at 4:06 A.M., then $m =$

21.

22. If x is a positive integer, the remainder when 2018^x is divided by 10 could NOT be
A) 4 B) 6 C) 8 D) 0

22.

Go on to the next page ⟹ A

23. If $a + b = 8$ and $\frac{1}{a} + \frac{1}{b} = 4$, then $ab =$ | 23.

A) 2 B) 6 C) 12 D) 32

24. Captain Fantabulous flies a route around the world 3 times, first at 900 km/hr., then at 1200 km/hr., then at 1800 km/hr. His average rate for all three laps around the world combined is _?_ km/hr. | 24.

A) 1200 B) 1300
C) 1400 D) 1500

25. The sum of all positive integers from 1 through $2n$ is | 25.

A) $2n^2 + n$ B) $2n^2 + 1$ C) $2n^2 + n + 1$ D) $2n^2 + 2n + 2$

26. If $3p - 4q = 5$ and $2p + 7q = 11$, then what is the value of $2p + 36q$? | 26.

A) 6 B) 16 C) 34 D) 55

27. There are 9 siblings in the Nunes family, and the sum of their ages is 110. If the average age of the 5 youngest siblings is 6, and the average age of the 5 oldest siblings is 18, what is the difference between the average age of the 4 oldest siblings and the average age of the 4 youngest siblings? | 27.

A) 8 B) 10 C) 15 D) 19

28. If $a - 3 = b + 6$, then $b^2 + 12b + 24 =$ | 28.

A) $a^2 - 6a - 3$ B) $a^2 - 6a + 9$ C) $a^2 + 6a - 12$ D) $a^2 + 6a$

29. Each year Mr. Lupo hires n animals; some are goats, and the rest are sheep. Two years ago, Mr. Lupo hired 20 more goats than sheep. Last year, Mr Lupo hired 6 more goats than he had hired two years ago. If goats were 3/4 of the animals hired last year, what is the value of n? | 29.

A) 18 B) 22 C) 48 D) 64

30. If x is the least positive integer with exactly 30 positive integer divisors, and s is the sum of the digits of x, then what is the value of s? | 30.

A) 9 B) 16 C) 25 D) 31

The end of the contest 

Visit our Web site at http://www.mathleague.com
Solutions on Page 121 • Answers on Page 149

Math League Press, P.O. Box 17, Tenafly, New Jersey 07670-0017

2018-2019 Annual Algebra Course 1 Contest

A

Spring, 2019

Instructions

- **Time** Do *not* open this booklet until you are told by your teacher to begin. You will have only *30 minutes* working time for this contest. You might be *unable* to finish all 30 questions in the time allowed.
- **Scores** Please remember that *this is a contest, and not a test*—there is no "passing" or "failing" score. Few students score as high as 24 points (80% correct). Students with half that, 12 points, *should be commended!*
- **Format and Point Value** This is a multiple-choice contest. Each answer is an A, B, C, or D. Write each answer in the *Answer Column* to the right of each question. A correct answer is worth 1 point. Unanswered questions receive no credit. You **may** use a calculator.

2018-2019 ALGEBRA COURSE 1 CONTEST

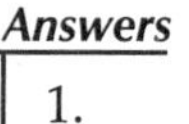
Answers

1. If $a = 2$, $r = 0$, $t = 1$, and $s = 9$, then $s + t + a + r + t =$

   A) 0 B) 12 C) 13 D) 21

   1.

2. There were a ants in my ant farm, but then 3 ants escaped! If each ant has 6 legs, the ants remaining have a combined total of _?_ legs.

   A) $6a - 3$ B) $6(a - 3)$ C) $6a - 3a$ D) $a^6 - 3$

   2.

3. $6x^2 - 5 + 4x - 3 + 2x^2 - 1 + 2x - 3 + 4x^2 - 5 + 6x =$

   A) $36x - 17$ B) $24x - 9$
   C) $12x^2 + 12x - 12$ D) $12x^2 + 12x - 17$

   3.

4. $(x - y)(x + y) =$

   A) $x^2 - y^2$ B) $x^2 - 2xy + y^2$ C) $x^2 + 2xy + y^2$ D) $x^2 + y^2$

   4.

5. $(x - y)(x + y)(x - y) =$

   A) $x^3 - y^3$ B) $x^3 - x^2y - xy^2 + y^3$
   C) $x^3 + y^3$ D) $x^3 + x^2y + xy^2 + y^3$

   5.

6. Which of the following is negative for all real values of s?

   A) $-s^3 - 1$ B) $(-s)^3 - 1$ C) $-s^2 - 1$ D) $(-s)^2 - 1$

   6.

7. $(x^2 - 1)(x^2 - 2)(x^2 - 3)(x^2 - 4) = 0$ has how many integer solutions?

   A) 2 B) 4 C) 6 D) 8

   7.

8. If x, y, and z are distinct prime numbers, which of the following is the least common multiple of $x^2y^3z^4$ and $x^4y^3z^2$?

   A) $x^8y^9z^8$ B) $x^6y^6z^6$ C) $x^4y^3z^4$ D) $x^2y^3z^2$

   8.

9. $((x^3 + x^3) \times x^3)^3 =$

   A) $2x^{18}$ B) $8x^{18}$ C) $8x^{27}$ D) x^{54}

   9.

10. In my big jar of jellybeans there are exactly $3b$ red beans, $5b$ green beans, and $6b$ orange beans, and no others. There could be a total of _?_ beans.

    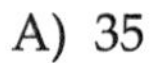
    A) 35 B) 42 C) 60 D) 90

    10.

11. What is the sum of all solutions to $|2x - 2.5| = 4$?

    A) 2 B) 2.5 C) 3.75 D) 4

    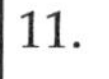
    11.

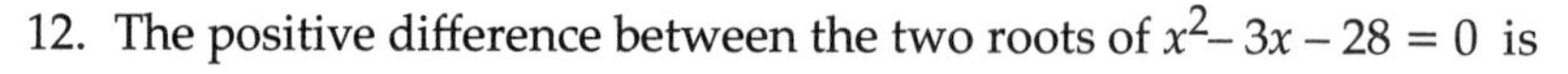

12. The positive difference between the two roots of $x^2 - 3x - 28 = 0$ is

    A) 3 B) 4 C) 7 D) 11

    12.

Go on to the next page ⟫⟫➔ A

	Answers
13. Today Li turned 42 and Mae turned 8. How old will Mae be when Li is exactly three times Mae's age? A) 9 B) 17 C) 26 D) 51	13.
14. If a crate of lightbulbs contains b boxes, and each box contains p packages, how many bulbs are in 3 crates if each package holds 4 bulbs? A) $12bp$ B) $\frac{3bp}{4}$ C) $\frac{4bp}{3}$ D) $\frac{bp}{12}$	14.
15. Avi and Bea were building sand castles all day. Avi had built three times as many castles as Bea, but then a wave destroyed 3 of Avi's castles while Bea built 1 more. At that point the ratio of Avi's castles to Bea's was 5:2. Avi had built __?__ castles before the wave hit. A) 11 B) 12 C) 30 D) 33	15.
16. If $135 \times 46 = a$, then $135 \times 48 =$ A) $a+2$ B) $a+92$ C) $a+94$ D) $a+270$	16.
17. If $3x + 8y = 21$ and $8x + 3y = 23$, then $x + y =$ A) 2 B) 4 C) 11 D) 22	17.
18. If the hands on a circular clock start at midnight, what number will the hour hand point to 1000 hours later? A) 2 B) 4 C) 8 D) 12	18.
19. If x is an integer, what is the least possible value of $\|20-7x\|$? A) 1 B) 2 C) 3 D) 6	19.
20. If Sy can shovel snow from half of a driveway in 2 hours, and Ty can shovel snow from one quarter of the driveway in 2 hours, how many *minutes* would it take them to shovel the whole driveway working together at their respective constant rates? A) 120 B) 160 C) 180 D) 360	20.
21. Of the bottles that Viola collects, 80% are green. Of the green bottles, 30% held perfume and 45% held spices. If the remaining 25 green bottles held pills, How many bottles are in Viola's collection? A) 75 B) 100 C) 120 D) 125	21.
22. If $x \neq 0$ and $2x - \frac{y-3x^2}{x} = \frac{4}{x}$, then $y =$ A) $4-x^2$ B) $4+x^2$ C) $5x^2-4$ D) $4-5x^2$	22.

Go on to the next page ⟹ **A**

23. Don and Juan had a total of x cherries, but then Don ate 27 fewer than x cherries and Juan ate 11 fewer than x cherries. If they each ate at least 10 cherries, and there was at least one cherry that wasn't eaten, then x =

A) 37 B) 38 C) 39 D) 49

23.

24. Of the 200 pets for sale at Pip's Pets, a have scales, b have gills, and c have both. How many of the pets have neither scales nor gills?

A) $200 - a - b$ B) $200 - c$ C) $200 - a - b - c$ D) $200 - a - b + c$

24.

25. The product of two numbers is 144, and the lesser of the two is 6 less than three times the greater. What is the greater of the two numbers?

A) 18 B) 8 C) -6 D) -24

25.

26. If x and y are positive numbers and $x + y = 2$, which of the following could be the value of $20x + 50y$?

A) 35 B) 65 C) 105 D) 140

26.

27. Iko's rectangular vegetable garden is $2x$ m wide and $3x$ m long. She wants to plant flowers to form a border of uniform width around the vegetable garden, and measures that the border will cover $14x^2$ m^2. How wide is the border of flowers going to be?

A) $0.5x$ m B) x m C) $1.5x$ m D) $2x$ m

27.

28. If $10^{2019} - 2019$ is written as an integer in decimal form, what is the sum of its digits?

A) 2019 B) 18160 C) 18161 D) 18169

28.

29. Tom mixes x kg of cake mix that is 10% sugar with y kg of cake mix that is 20% sugar. If the resulting mixture is z% sugar, then the ratio of x to y is

A) $(20 - z):(z - 10)$ B) $(10 - z):(z + 20)$
C) $(z + 10):(20 - z)$ D) $(z + 20):(10 - z)$

29.

30. If x, y and z are prime, what is the product of all whole-number divisors of the product xyz?

A) xyz B) $x^2y^2z^2$ C) $x^3y^3z^3$ D) $x^4y^4z^4$

30.

The end of the contest  **A**

Visit our Web site at http://www.mathleague.com

Solutions on Page 125 • Answers on Page 150

Math League Press, P.O. Box 17, Tenafly, New Jersey 07670-0017

2019-2020 Annual Algebra Course 1 Contest

A

Spring, 2020

Instructions

- **Time** Do *not* open this booklet until you are told by your teacher to begin. You will have only *30 minutes* working time for this contest. You might be *unable* to finish all 30 questions in the time allowed.
- **Scores** Please remember that *this is a contest, and not a test*—there is no "passing" or "failing" score. Few students score as high as 24 points (80% correct). Students with half that, 12 points, *should be commended!*
- **Format and Point Value** This is a multiple-choice contest. Each answer is an A, B, C, or D. Write each answer in the *Answer Column* to the right of each question. A correct answer is worth 1 point. Unanswered questions receive no credit. You **may** use a calculator.

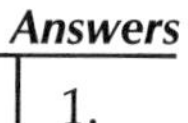

Answers

1. If $T = 1$, $O = 2$, and $T+O+S+S = 7$, then $S =$

   A) 2 B) 3 C) 3.5 D) 4

2. If x is an integer, then the least possible value of $4x^2$ is

   A) -4 B) 0 C) 4 D) 16

3. $(c^{20})(c^2)(c^0) =$

   A) 0 B) c^0 C) c^{22} D) c^{40}

4. I had g invited guests at my party. Each invited guest brought 2 uninvited friends. Each person who came, whether invited or not, brought two gifts. How many gifts were brought?

   A) $(g+2)\times 2$ B) $(g\times 2)+2$ C) $(g+2g)\times 2$ D) $(g+2g)\times 2g$

5. $4y(x-y)-(3x+2y)(x-y) =$

   A) $(6y+3x)(x-y)$ B) $(6y-3x)(x-y)$
   C) $(2y+3x)(x-y)$ D) $(2y-3x)(x-y)$

6. $4x^2+3x+2x^3-2x^2-3x-4x^3 =$

   A) 0 B) $2x^2-2x^3$ C) $2x^2+6x-2x^3$ D) $2x^2+6x+6x^3$

7. If $\frac{3}{5}$ of $2y$ is equal to $\frac{4}{7}$ of x, then what is y in terms of x?

   A) $\frac{10}{21}x$ B) $\frac{20}{21}x$ C) $\frac{21}{20}x$ D) $\frac{21}{10}x$

8. How many distinct solutions does $(x+2)(x-2)(x^2-4) = 0$ have?

   A) 1 B) 2 C) 3 D) 4

9. If $x > 5$ and x is prime, the least common multiple of $20x^2$ and $30x^3$ is

   A) $10x$ B) $60x^3$ C) $60x^5$ D) $600x^5$

10. For every 30 sec. Peg and her dog Al are in the the water, Peg later walks Al for 2 min. If they spent a total of h hrs. combined in the water and walking, they spent __?__ min. in the water.

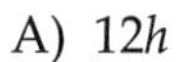

    A) $12h$ B) $24h$ C) $36h$ D) $48h$

11. If $x-y = 1$ and $x^2-y^2 = 39$, then $xy =$

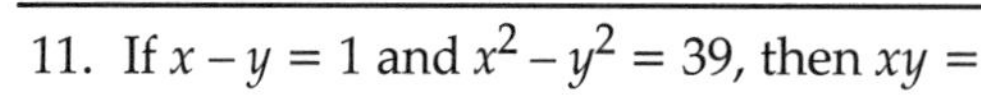

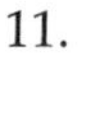

    A) 39 B) 78 C) 380 D) 1521

12. What is the remainder when x^3-x^2+x-1 is divided by $x-1$?

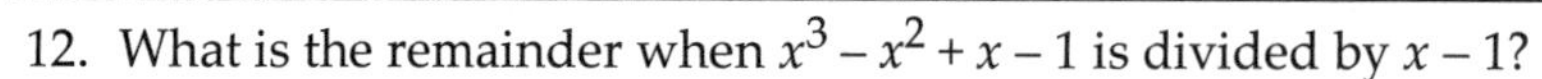

    A) 0 B) 1 C) x D) $2x$

Go on to the next page ⟹ A

13. What is the equation of a line perpendicular to $y = \frac{1}{3}x + 4$ and with the same x-intercept?

A) $y = -3x + 4$ B) $y = 3x - 36$ C) $y = \frac{1}{3}x + 4$ D) $y = -3x - 36$

13.

14. What is the sum of both solutions to $4x^2 - 4x - 35 = 0$?

A) -1 B) 0 C) 1 D) 4

14.

15. Someone replaced Emma's g eggs with b billiard balls! Emma notices that $b + 1 = g^2$ and $b^2 + 31 = g^4$. What is the value of g?

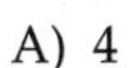

A) 4 B) 11 C) 12 D) 30

15.

16. Emma flies k km at r m/min. to retrieve her missing eggs. She flies for __?__ hours.

A) $\frac{1000}{rk}$ B) $\frac{60k}{1000r}$ C) $\frac{1000k}{60r}$ D) $\frac{1000}{60rk}$

16.

17. $4^{2x} + 4^{2x} + 4^{2x} + 4^{2x} =$

A) 2^{4x} B) 2^{4x+2} C) 4^{8x} D) 16^{2x}

17.

18. If $f(x) = 8x^2 - 2$, then which of the following is equal to $f(4)$?

A) $f(126)$ B) $f(8)$ C) $f(-2)$ D) $f(-4)$

18.

19. How many integer values of x satisfy $|3x - 7| < 5$?

A) 1 B) 2 C) 3 D) 6

19.

20. If $x^2 + x + 1 = 18$, then the average of x^3, x^2, and x is

A) $6x$ B) $9x$ C) $18x$ D) $36x$

20.

21. The Cones is an elite *a cappella* vocal group. To be accepted, a Cone has to be good at setting up chairs in the theater and have a great voice. Together, four Cones working at the same rate can set up every chair in the theater in 56 min. At this same rate, it would take __?__ min. for 7 Cones to set up every chair.

A) 24 B) 32 C) 48 D) 98

21.

22. If 10^a is 0.01 percent of 10^b, then $a =$

A) $b - 4$ B) $b - 2$ C) $b + 2$ D) $b + 4$

22.

Go on to the next page ⟹ A

Question	Answers
23. Ishmael hooked a big fish! The fish dragged his boat x km east, then $6x + 3$ km north, then $x + 8$ km east, then $2x - 2$ km south, then $2x + 8$ km west, then x^2 km south, $x > 0$. Then, the fish escaped. Ishmael looked around to find that he was exactly where he had started. How many km long was the route taken by the fish and Ishmael? A) 5 B) 30 C) 81 D) 102  	23.
24. Last week I mixed up 400 ml of lemonade that was 30% sugar and 200 ml of lemonade that was 40% sugar, and then poured all of it into a glass. During the week 100 ml of pure water evaporated from the lemonade. What percent sugar is the remaining lemonade? A) 30% B) 33% C) 35% D) 40%	24.
25. If $x \neq 0$, $y \neq 0$, and $\sqrt{xy} \times \sqrt{15} = \sqrt{3x^2} \times \sqrt{y}$, then $x =$ A) 5 B) y C) $5y$ D) $5 + y$	25.
26. If $x \neq 6$ and $x \neq -3$, then $\frac{x^2 - 3x - 18 - x + 6}{(x-6)(x+3)} =$ A) 1 B) $\frac{x+2}{x+3}$ C) $\frac{x-6}{x+3}$ D) $\frac{-2}{-(x-6)}$	26.
27. If $3x - 4y + 5z = 13$ and $4x - 5y + 6z = 18$, then $x - 2y + 3z =$ A) 1 B) 3 C) 15 D) 28	27.
28. Sir Saul was the sole survivor of the siege. There were between 1000 and 5000 knights at the start, but each day, two-thirds of the remaining knights fell or fled. Yesterday Saul lost his final 2 fellow knights. How many days ago did the siege start? A) 70 B) 14 C) 7 D) 6 	28.
29. If $8^{2a} = 32b$, then $b =$ 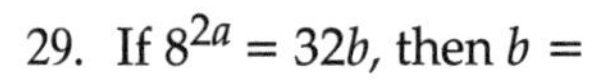 A) 2^a B) $2^{6a/5}$ C) 2^{2a-3} D) 2^{6a-5}	29.
30. If p is the product of all integers from 1 through 1000, what is the greatest integer q such that 4^q is a factor of p? A) 250 B) 312 C) 330 D) 497	30.

The end of the contest  **A**

Visit our Website at http://www.mathleague.com

Solutions on Page 129 • Answers on Page 151

Math League Press, P.O. Box 17, Tenafly, New Jersey 07670-0017

2020-2021 Annual Algebra Course 1 Contest

Spring, 2021

Instructions

- **Time** Do *not* open this booklet until you are told by your teacher to begin. You will have only *30 minutes* working time for this contest. You might be *unable* to finish all 30 questions in the time allowed.
- **Scores** Please remember that *this is a contest, and not a test*—there is no "passing" or "failing" score. Few students score as high as 24 points (80% correct). Students with half that, 12 points, *should be commended!*
- **Format and Point Value** This is a multiple-choice contest. Each answer is an A, B, C, or D. Write each answer in the *Answer Column* to the right of each question. A correct answer is worth 1 point. Unanswered

2020-2021 ALGEBRA COURSE 1 CONTEST

Answers

1. If $x = -2021$, which of the following is greatest?

   A) $-2021x$ B) $-2020x$ C) $2020x$ D) $2021x$

2. The letters in the 9-letter code *MATHISFUN* represent the 9 digits 1, 2, . . . , 9 in that order. What is the 4-digit number represented by the code *TINA*?

   A) 3492 B) 3582 C) 3592 D) 3594

3. If $2x + 5 = 13$, what is the value of $8x + 20$?

   A) 26 B) 29 C) 39 D) 52

4. If $8a = 12b$ and $ab \neq 0$, what is the value of $\frac{a}{b}$?

   A) $\frac{1}{2}$ B) $\frac{2}{3}$ C) $\frac{3}{2}$ D) 2

5. $2a - 3b + 4 - 9a + 13b - 15 =$

   A) $-7a - 10b - 11$ B) $-7a + 10b - 11$
   C) $7a + 10b - 11$ D) $-7a + 10b + 11$

6. $(a + 2)(3a - 4) =$

   A) $a^2 - 2a - 8$ B) $3a^2 - 2a - 8$ C) $3a^2 - 2a + 8$ D) $3a^2 + 2a - 8$

7. $x^2 - (x + 4)(x - 4) =$

   A) -16 B) -8 C) 8 D) 16

8. For all integers x, $125x^6$ is **always** the __?__ power of an integer.

   A) 2nd B) 3rd C) 5th D) 6th

9. The sum of the ages of my 3 children is 45 years. If their ages are consecutive odd integers, how many years old is my oldest child?

   A) 17 B) 19 C) 21 D) 23

10. If $a - b = 10$ and $ab = 50$, then $\frac{1}{a} - \frac{1}{b} =$

    A) -5 B) $-\frac{1}{5}$ C) $\frac{1}{5}$ D) 5

11. If a and b are positive numbers, then $\sqrt{2.56a^6b^{10}} =$

    A) $1.28a^3b^5$ B) $1.28a^6b^{10}$ C) $1.6a^3b^5$ D) $1.6a^6b^{10}$

12. If today is Tuesday, $14n + 3$ days from now (where n is a positive integer) will be a

    A) Wednesday B) Thursday C) Friday D) Saturday

Go on to the next page **A**

13. In my school, 4% of the students are vegan. If 120 students are not vegan, how many students are in my school?

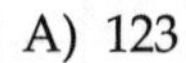

A) 123 B) 124 C) 125 D) 126

13.

14. If $5 < m < 10$, what is the median of the following numbers: $m + 5$, $m + 10$, $m + 14$, $m + 20$, $m + 22$, and $2m$?

A) $m + 12$ B) $m + 17$ C) $m + 21$ D) $(3m + 14)/2$

14.

15. The total number of points of intersection of the graphs of $4x + 10y = 12$ and $6x + 15y = 20$ is

A) 0 B) 1 C) 2 D) 3

15.

16. A large bundle of stamps has twice as many stamps as a small bundle. Jack has 5 large bundles of stamps and 7 individual stamps, while Jill has 11 small bundles of stamps and 2 individual stamps. Jack has the same number of stamps as Jill. How many stamps does each have?

A) 57 B) 67 C) 68 D) 78

16.

17. What is the remainder when $x^4 + x^3 + x^2 + x + 1$ is divided by $x - 1$?

A) 1 B) 5 C) x D) $5x$

17.

18. If $x \bigstar y = xy(x - 3y)$, what is the value of $(2 \bigstar 1) \bigstar (4 \bigstar 1)$?

A) -112 B) -54 C) 54 D) 112

18.

19. How many different real numbers x satisfy $|3x + 5| = 2x$?

A) 0 B) 1 C) 2 D) 3

19.

20. What is the positive value of x that satisfies $x^{400} = 9^{1000}$?

A) 27 B) 81 C) 243 D) 729

20.

21. The number of hot dogs I ate at the picnic is the same as the number of real numbers that satisfy $(x^2 - 12)^2 = 169$. How many hot dogs did I eat?

A) 1 B) 2 C) 3 D) 4

21.

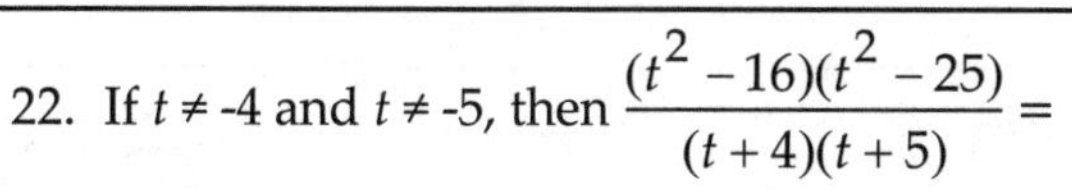

22. If $t \neq -4$ and $t \neq -5$, then $\dfrac{(t^2 - 16)(t^2 - 25)}{(t + 4)(t + 5)} =$

A) $(t + 4)(t + 5)$ B) $(t - 4)(t - 5)$ C) $(t + 16)(t + 25)$ D) $(t - 16)(t - 25)$

22.

Go on to the next page ➠ **A**

23. For how many different values of x is $\frac{12}{x}$ an integer?

A) 5 B) 6 C) 12 D) more than 12

23.

24. What is the value of b for which $x^2 - 8x + b = 0$ has only one solution?

A) 64 B) 32 C) 16 D) 8

24.

25. Working together, 5 printers can print 640 pages in 4 minutes. Working at this same rate, how many minutes would it take 8 printers to print 2560 pages?

A) 8 B) 10 C) 12 D) 16

25.

26. If n is a positive integer, $n(n+1)(n+2)(n+3)$ must be divisible by which of the following numbers?

A) 8 B) 9 C) 10 D) 14

26.

27. I can buy b bags of chips from a vending machine for m dimes. If I have d dollars, at most how many bags of chips can I buy?

A) $\frac{10dm}{b}$ B) $\frac{10bd}{m}$ C) $\frac{10m}{bd}$ D) $\frac{bd}{10m}$

27.

28. On Monday a jacket was marked down by 65% from its original price. On Tuesday the jacket was marked down by 65% from Monday's discounted price, an additional $91 decrease. What was the jacket's original price before both discounts?

A) $400 B) $450 C) $500 D) $550

28.

29. If $3^x = 2$ and $27^{2x+1} = 32a$, then $a =$

A) 18 B) 27 C) 36 D) 54

29.

30. If the average of x and y is 15, the average of y and z is 24, and the average of x and z is 27, what is the average of x, y, and z?

A) 22 B) 28 C) 36 D) 44

30.

The end of the contest A

Visit our Website at http://www.mathleague.com

Solutions on Page 133 • Answers on Page 152

Detailed Solutions

2016-2017 through 2020-2021

7th Grade Solutions

2016-2017 through 2020-2021

Math League Press, P.O. Box 17, Tenafly, New Jersey 07670-0017

Information & Solutions

7

Tuesday, February 21 or 28, 2017

Contest Information

- **Solutions** Turn the page for detailed contest solutions (written in the question boxes) and letter answers (written in the *Answer Column* to the right of each question).

- **Scores** Please remember that *this is a contest, and not a test*—there is no "passing" or "failing" score. Few students score as high as 28 points (80% correct); students with half that, 14 points, *deserve commendation!*

- **Answers and Rating Scales** Turn to page 138 for the letter answers to each question and the rating scale for this contest.

	Answers
1. $12 \div -4 = -3$. A) $-\frac{1}{3}$ B) -3 C) 3 D) $\frac{1}{3}$	1. B
2. The ratio of pens to pencils on my desk is 5:8. There could be 10 pens and 16 pencils on my desk. A) 10 B) 12 C) 14 D) 16	2. D
3. The product of 5.96 and 3.06 is 18.2376. Rounded to the nearest whole number, this is 18. A) 15 B) 16 C) 18 D) 20	3. C
4. $(77-66)+(55-44)+(33-22)+11 = 44$. A) 33 B) 44 C) 77 D) 110	4. B
5. The smaller a number is, the smaller its square root is. A) $\sqrt{\frac{2}{3}}$ B) $\sqrt{\frac{3}{4}}$ C) $\sqrt{\frac{4}{5}}$ D) $\sqrt{\frac{5}{6}}$	5. A
6. Since 5 is not a factor of 4, 6, or 8, 30 is not a factor of $4 \times 6 \times 8$. A) 16 B) 24 C) 30 D) 96	6. C
7. The ages of my 3 cousins and my 3 uncles are 14, 15, 16, 42, 45, and 48. Their average age is $(14+15+16+42+45+48) \div 6 = 30$. A) 30 B) 29 C) 28 D) 27	7. A
8. $4-2 \div \left(\frac{1}{2}\right) = 4-4 = 0$. A) 0 B) 3 C) 3.75 D) 4	8. A
9. My field's perimeter is $2\times(80+40)$ m = 240 m and its area is 3200m^2. A square with this perimeter has sides of length 240 m $\div$ 4 = 60 m. The square's area is 3600 m^2. The areas differ by 400 m^2. A) 200 m^2 B) 400 m^2 C) 600 m^2 D) 800 m^2	9. B
10. Since 6 dimes + 2 nickels = 70¢, 5 dimes + 3 nickels = 65¢, 3 dimes + 5 pennies = 35¢, choice D is correct. A) 70¢ B) 65¢ C) 35¢ D) 15¢	10. D
11. $(21\times2\times27\times2\times29\times2\times23\times2) \div (29\times21\times23\times27) = 2\times8$. A) 2 B) 4 C) 6 D) 8	11. D
12. Pick any number and its reciprocal. Let's use 3 and 1/3. Multiply 3 by 4 to get 12. Its reciprocal is 1/12, so we must multiply 1/3 by 1/4 to get 1/12. A) -4 B) 1 C) $-\frac{1}{4}$ D) $\frac{1}{4}$	12. D
13. Opposite sides of a parallelogram, rhombus, and square are equal. A) parallelogram B) rhombus C) trapezoid D) square	13. C

Go on to the next page ⟹ 7

Answers

14. Using a common denominator of 24, the average is

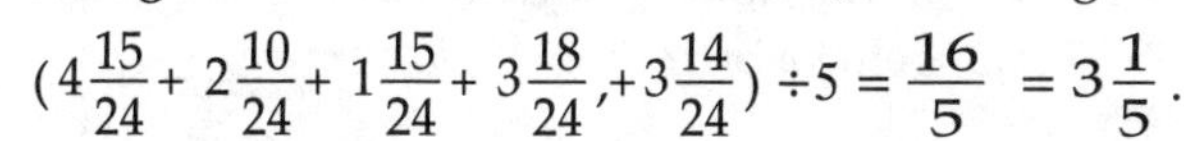

$(4\frac{15}{24} + 2\frac{10}{24} + 1\frac{15}{24} + 3\frac{18}{24}, + 3\frac{14}{24}) \div 5 = \frac{16}{5} = 3\frac{1}{5}$.

A) $3\frac{1}{4}$ B) $3\frac{1}{3}$ C) $3\frac{1}{5}$ D) $3\frac{1}{2}$

14. C

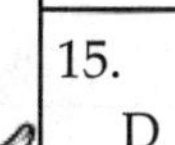

15. Use 2, 4, 6, or 8 in the ten thousands place, then the remaining digits in the other places: 4×4×3×2×1 = 96.

A) 625 B) 384 C) 120 D) 96

15. D

16. $2^{2017} - 2^{2016} = 2^{2016} \times (2-1) = 2^{2016}$.

A) 1 B) 2 C) 2^{2016} D) 2^{2017}

16. C

17. As shown below, the largest number is 5, which is choice A.

A) $\frac{-2-3}{-1} = 5$ B) $\frac{12-4}{3} = 8/3$ C) $\frac{12+3}{5} = 3$ D) $\frac{8-2}{-1} = -6$

17. A

18. The least number of different single-digit numbers I can choose whose product is greater than 2017 is 4, since 9×8×7×6 is 3024.

A) 3 B) 4 C) 6 D) 7

18. B

19. When 7677 is divided by 7, the remainder is 5.

A) 7677 B) 7686 C) 7714 D) 7777

19. A

20. If there are 8 nits in 1 nat, there are $8 \times 6.25 = 50$ nits in 6.25 nats.

A) 49 B) 50 C) 51 D) 52

20. B

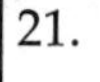

21. A ratio of 4:10 equals 2:5 or 6:15 or . . .. Find the ratio that yields a 1:2 ratio when 3 is added to each number; 6:15 works, so 18 is the new total.

A) 12 B) 15 C) 16 D) 18

21. D

22. Powers of 5 > 1 end in 25; 25÷4 has remainder 1.

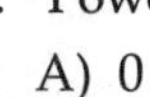

A) 0 B) 1 C) 2 D) 3

22. B

23. Choose one of the odd numbers and the only even number, 80. There are 4 odd numbers, so there are 4 ways to do this.

A) 4 B) 8 C) 24 D) 60

23. A

24. The least value occurs when the numbers are 50 and 50. The sum of their reciprocals is 1/50 + 1/50 = 2/50 = 4/100 = 0.04.

A) 0.01 B) 0.04 C) 0.05 D) 0.1

24. B

25. Choice B is correct since $(-0.5)^3 = -0.625 < (-0.5)^2 = 0.25$.

A) $(-0.5)^3 > (-0.5)^2$ B) $(-0.5)^3 < (-0.5)^2$

C) $(-0.5)^3 = (-0.5)^2$ D) $(-0.5)^3 = -(-0.5)^2$

25. B

Go on to the next page ⟹ 7

Answers

26. Fit 7 of the 13 cm edges on the 91 cm edge, and 13 of the 10 cm edges on the 130 cm side. That's a 7×13 arrangement of 91 tiles.

A) 90 B) 91 C) 117 D) 130

26. B

27. The ratio of $3 : \frac{1}{2}$ is 6. The next term is 3 × 6 = 18.

A) 6 B) 12 C) 18 D) 36

27. C

28. At Burgertown I can buy 4 hot dogs and 5 burgers for \$60, or 4 hot dogs and 6 burgers for \$66. The cost of 1 burger is \$66 − \$60 = \$6.

A) \$6.00 B) \$6.50 C) \$7.00 D) \$7.50

28. A

29. $99 \times 665 = 3^2 \times 11 \times 5 \times 7 \times 19$
$= 3 \times 7 \times 11 \times (3 \times 5) \times 19$

A) 6×9×11×65 B) 100×665 − 99
C) 3×7×11×15×19 D) 7×11×15×19×23

29. C

30. (99 − 98) + (97 − 96) + (95 − 94) + . . . + (5 − 4) + (3 − 2) + 1 = 49 + 1 = 50.

A) 49 B) 50 C) 99 D) 148

30. B

31. There are 3600 seconds in an hour; 30/3600 = 1/120.

A) $\frac{1}{2}$ B) $\frac{1}{60}$ C) $\frac{1}{120}$ D) $\frac{1}{240}$

31. C

32. He sips 10 ml of medicine on the 1st day. After replacing this 10 ml with water, his vial now has 70 ml of medicine in it. He sips 15 ml from the vial the 2nd day. Since 70/80 is medicine, he sips 7/8 × 15 ml = 105/8 ml of medicine. This is 25/8 ml more than on the 1st day.

A) $\frac{5}{2}$ ml B) $\frac{25}{8}$ ml C) $\frac{35}{8}$ ml D) 5 ml

32. B

33. The sum of the perimeters of the two congruent triangles is equal to the perimeter of the original triangle plus twice the altitude. Thus, 76 − 48 = 28 is twice the altitude; the altitude is 14.

A) 14 B) 19 C) 24 D) 38

33. A

34. 366÷7 = 52 R2; if Jan. 1 is a Sun. or Mon., there are 53 Mondays.

A) 0 B) 1/7 C) 2/7 D) 3/7

34. C

35. Since $20^{17} = 2^{17} \times 10^{17} = 131\,072 \times 10^{17}$, it has 6 + 17 = 23 digits.

A) 32 B) 28 C) 25 D) 23

35. D

The end of the contest ✍ 7

Visit our Website at http://www.mathleague.com

Information & Solutions

7

Tuesday, February 20 or 27, 2018

Contest Information

- **Solutions** Turn the page for detailed contest solutions (written in the question boxes) and letter answers (written in the *Answer Column* to the right of each question).

- **Scores** Please remember that *this is a contest, and not a test*—there is no "passing" or "failing" score. Few students score as high as 28 points (80% correct); students with half that, 14 points, *deserve commendation!*

- **Answers and Rating Scales** Turn to page 139 for the letter answers to each question and the rating scale for this contest.

	Answers
1. If you start with 2 and double it twice, you get 8. If you now triple 8, you get 24. This is 1200% of your starting integer. A) 700% B) 1100% C) 1200% D) 1300%	1. C
2. Barry listened to the radio for 3 hours and 36 minutes, which is 216 minutes. Rounded to the nearest 10 minutes, this is 220 minutes. A) 210 B) 220 C) 330 D) 340	2. B
3. Divide 99 by 22 to get a quotient of 4 and a remainder of 11. Now divide 11 by 4; the quotient is 2 and the remainder is 3. A) 4 B) 3 C) 2 D) 1	3. B
4. $10 - [(8 \div 4) \times 6] + 2 = 10 - 12 + 2 = 0$. A) 0 B) 2 C) 5 D) 12	4. A
5. The product of 2, 5, 10, and 20 is 2000, so the hundreds digit is 0. A) 6 B) 4 C) 2 D) 0	5. D
6. The sum of the prime factors of 2018 is $2 + 1009 = 1011$. A) 11 B) 219 C) 1011 D) 2019	6. C
7. A triangle has one side of length 18. The sum of the lengths of the other two sides must be more than 18, so at least one has length > 9. A) 9 B) 10 C) 12 D) 17	7. A
8. My final score is the average of my scores on 5 rounds. If my final average of 5 rounds is 88, my total is $5 \times 88 = 440$. After getting $84 + 80 + 92 = 256$, my last 2 rounds must total 184. That is a 92 average. A) 88 B) 90 C) 92 D) 96	8. C
9. There are 71 integers between 19 and 91; 36 are even. A) 35 B) 36 C) 37 D) 71	9. B
10. The ratio of freshmen to other students in Prof. Peach's class is 3:8, so the ratio of freshman to all students is 3:(3+8). The total number of students must be divisible by 11. A) 42 B) 45 C) 56 D) 77	10. D
11. $4^{40} \div 2^{20} = (2^2)^{40} \div 2^{20} = 2^{80} \div 2^{20} = 2^{60}$ A) 2^2 B) 2^4 C) 2^{20} D) 2^{60}	11. D
12. $2 \times 12 + 6 + 12 = 24 + 6 + 12 = 42$. A) 32 B) 36 C) 42 D) 48	12. C
13. \$5 – (5 quarters + 5 dimes + 5 nickels) = \$5 – (\$1.25+\$0.50+\$0.25) = \$3. A) \$3.00 B) \$3.25 C) \$3.45 D) \$3.75	13. A

Go on to the next age 7

14. One side of Todd's truck is a rectangle with an area of $12\ m^2$. If its length is 3 times its width, its dimensions are 2 m and 6 m.

    A) 8 m B) 12 m C) 16 m D) 20 m

    14. C

15. If a bird in the hand is worth 2 in the bush, and a bird in the bush is worth 4 in the sky, then 4 birds in the hand are worth 8 in the bush and 32 in the sky.

    A) 1 B) 4 C) 16 D) 32

    15. D

16. The four shelves of my bookcase could hold 2, 3, 5, and 11 books. There could be a total of $2 + 3 + 5 + 11 = 21$ books on my shelves.

    A) 15 B) 21 C) 22 D) 24

    16. B

17. If 12 years passing triples my age, then 12 is twice my starting age, which would be 6. If I was 6 seven years ago, I am now 13.

    A) 11 B) 13 C) 16 D) 18

    17. B

18. The fractions between 1 and 1/100 with a numerator of 1 and a whole-number denominator are 1/2, 1/3, . . . , 1/98, and 1/99.

    A) 98 B) 99 C) 100 D) 101

    18. A

19. Since $100 \div 6 = 16R4$, I write the entire word 16 times plus 4 additional letters, R-E-P-E. I write the letter "E" $2 \times 16 + 2 = 34$ times.

    A) 16 B) 18 C) 32 D) 34

    19. D

20. The area of the park on the map is $2.5\ cm \times 4\ cm = 10\ cm^2$. Since $1\ cm^2$ represents $10\,000\ km^2$, the actual area is $100\,000\ km^2$.

    A) 100 B) 1000 C) 10000 D) 100000

    20. D

21. Gus's cloud shows up at 8:30 A.M. and every 50 minutes after that. It 1st shows up 930 minutes before midnight; it appears 30 minutes before 12 A.M.

    A) 11:00 B) 11:10 C) 11:30 D) 11:50

    21. C

22. If my lucky number divided by its reciprocal is 100, then my lucky number is 10 and its square is 100.

    A) 100 B) 10 C) 1 D) $\frac{1}{100}$

    22. A

23. Five consecutive even integers have a sum of 190. Their average is 38 and the largest is 4 more, 42.

    A) 34 B) 38 C) 40 D) 42

    23. D

24. The ratio 7.8 to 5.2 is equivalent to 78 to 52 or 3 to 2.

    A) 8 to 5 B) 7 to 5 C) 3 to 2 D) 7 to 2

    24. C

25. 15% of 80 = 12 = 40% of 30.

    A) 30 B) 55 C) 105 D) 210

    25. A

Go on to the next page ⟹ **7**

Solution	Answers
26. It took me 90 minutes to cycle 45 km to the beach, so it took me 2 minutes to cycle 1 km. My ride's speed was twice my cycling speed, so it took me 1 minute to ride 1 km. In 15 minutes, I traveled 15 km. A) 15 km B) 30 km C) 45 km D) 135 km	26. A
27. Suppose Rick painted 100 units. Then Sam painted 60 units and Ted painted 108 units. Therefore, Ted painted 108% of the amount that Rick painted. A) 102% B) 108% C) 120% D) 140%	27. B
28. $50^{50} = (2 \times 5^2)^{50} = 2^{50} \times 5^{100}$. Since $20 = 2^2 \times 5$, there are only enough 2s for 20^{25}. A) 20^{20} B) 20^{25} C) 20^{50} D) 20^{125}	28. B
29. The ones digit is 2 more than the ones digit of $149 \times (2+4+6+8+0)$. A) 2 B) 4 C) 8 D) 0	29. A
30. The median is the average of the middle two: $(\frac{1}{4} + \frac{1}{5}) \div 2 = \frac{9}{40}$. A) $\frac{1}{8}$ B) $\frac{223}{840}$ C) $\frac{5}{14}$ D) $\frac{9}{40}$	30. D
31. The average of consecutive even integers is the average of 1st and last. A) 1000 B) 1009 C) 1010 D) 1014	31. C
32. Pirate Percy has 300 coins in his chest. Since 100 are gold coins that are not Spanish, and 70 are neither Spanish nor gold, the remaining 130 coins are Spanish. If 20% of these 130 coins are gold, then there are 26 Spanish gold coins. A) 20 B) 26 C) 30 D) 34	32. B
33. Since 0.06 = 6/100 = 18/300, I must have divided the population of my city by 300. Therefore, the number of streets in my city is 300. A) 60 B) 120 C) 186 D) 300	33. D
34. Fit 11 rows of 30 small rectangles with the shorter side on the side of length 90 for 330 rectangles. Fit 12 more in the 3 by 90 rectangle left. A) 312 B) 330 C) 334 D) 342	34. D
35. If the thousands digit is a 6, 4, or 2, there are 1, 2, or 3 possibilities for the ones digit. The other digits may be chosen in 6 ways for a total of $6 \times (1+2+3.)$. A) 36 B) 54 C) 60 D) 90	35. A

The end of the contest 7

Visit our Website at http://www.mathleague.com

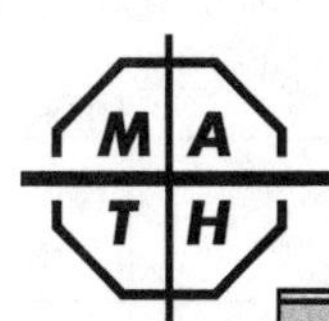

Math League Press, P.O. Box 17, Tenafly, New Jersey 07670-0017

Information & Solutions

7

Tuesday, February 19 or 26, 2019

Contest Information

- **Solutions** Turn the page for detailed contest solutions (written in the question boxes) and letter answers (written in the *Answer Column* to the right of each question).

- **Scores** Please remember that *this is a contest, and not a test*—there is no "passing" or "failing" score. Few students score as high as 28 points (80% correct); students with half that, 14 points, *deserve commendation!*

- **Answers and Rating Scales** Turn to page 140 for the letter answers to each question and the rating scale for this contest.

	Solution	Answers
1.	$(2 \times 4 \times 8) \div 2 = 1 \times 4 \times 8 = 4 \times 8$. A) 2 B) 4 C) 8 D) 16	1. C
2.	Al sleeps daily for 3/4 of the day. Therefore, Al sleeps for 18 hours. A) 6 B) 9 C) 12 D) 18	2. D
3.	$36 = 6 \times 6 = (-6) \times (-6)$. A) -6 B) 6 C) -30 D) 42	3. A
4.	$20 \times (19 - 1) = 20 \times 19 + 20 \times (-1)$. A) -1 B) 0 C) 1 D) 20	4. A
5.	Since $\frac{3}{10}$ of 60 minutes is 18 minutes, Angel arrived 18 minutes before 12 p.m.; that's 11:42 a.m. A) 11:18 a.m. B) 11:20 a.m. C) 11:40 a.m. D) 11:42 a.m.	5. D
6.	The product of 1 and 2019 is 2019. A) 673 B) 2019 C) 2020 D) 6057	6. B
7.	The sum of the first ten whole numbers is 45. Their average is 4.5. A) 5.5 B) 5 C) 4.5 D) 4	7. C
8.	$2019 \times 3 + 2019 \times 1/3 = 2019 \times (3 + 1/3)$. A) 0 B) $\frac{1}{3}$ C) 1 D) 3	8. B
9.	The product of four 4s = 256 = 4×64; this is the sum of 64 4s. A) 4 B) 3×4 C) 4^3 D) 4^4	9. C
10.	If 1/3 the side-length is 4, the side-length is 12 and the area is 144. A) 12 B) 16 C) 48 D) 144	10. D
11.	Doubling 20 six times, my surf club had 2560 members 7 days later. Seven days after a Monday is also a Monday. A) Sunday B) Monday C) Tuesday D) Friday	11. B
12.	A number such as 4.5 is rounded to 5, an increase of 0.5. This is the greatest possible increase when a number is rounded to the nearest integer. A) 0.05 B) 0.1 C) 0.5 D) 0.9	12. C
13.	The perimeter is greatest when the length is 2019 and the width is 1. The difference between dimensions is at most 2018. A) 0 B) 1 C) 670 D) 2018	13. D

Go on to the next page ⟹ 7

Solution	Answers
14. Using 6 pals, 3 pals have at least 1 pet, and $\frac{1}{3}$ of them, or 1 pal, has more than 1 pet. The fraction of my pals with exactly 1 pet is 2/6. A) $\frac{1}{6}$ B) $\frac{1}{3}$ C) $\frac{2}{3}$ D) $\frac{5}{6}$	14. B
15. The average of 0.5, 1.5, and 2.5 is 1.5; the average of 1 and 2 is also 1.5. A) 1 B) 1.5 C) 2 D) 2.5	15. C
16. $9 \times (9 \times 10) \times (9 \times 100) \times (9 \times 1000) = 9 \times (9^3 \times 1\,000\,000) = 9 \times 900^3$. A) 100^3 B) 900^3 C) 9000^3 D) $9\,000\,000^3$	16. B
17. The number one less than -342 is -343. A) -341 B) -342 C) -343 D) -344	17. C
18. The number of digits in the decimal form of 10^{2018} is 2019; 2019 ÷ 4 is 504R3. A) 3 B) 2 C) 1 D) 0	18. A
19. The number of letters in the first name "Ali" is 60% of the number of letters in a 5-letter last name. A) Al B) Ali C) Alex D) Alexa	19. B
20. $12 = \pm1 \times \pm12 = \pm2 \times \pm6 = \pm3 \times \pm4$; the least sum is $-1 + (-12) = -13$. A) -13 B) -11 C) 7 D) 8	20. A
21. Since 100 ÷ 6 = 16R4, 16 are multiples of both 2 and 3, and 84 are not. A) 16 B) 32 C) 64 D) 84	21. D
22. Since each carton contains 8 eggs that are not cracked, 3 cartons contain 2 dozen eggs that are not cracked. I need 24 cartons in all. A) 48 B) 36 C) 24 D) 20	22. C
23. In order, the choices are 8.40, 8.20, 8.50 and 8.40: 8.20 is nearest. A) $8\frac{2}{5}$ B) $8\frac{2}{10}$ C) $8\frac{5}{10}$ D) $8\frac{10}{25}$	23. B
24. Each day can be paired with 6 other days for a total of 42 pairs. However, each pair has been counted twice, so there are 21 pairs. A) 14 B) 21 C) 28 D) 35	24. B
25. Write with 4 digits to the right of the decimal. A) 0.1 B) 0.01 C) 0.0011 D) $(0.01)^2$	25. D

Go on to the next page ⟹ 7

26. Such a prism has 4 edges of each size. The sum of the 3 dimensions is 15 m, so the sum of all the lengths is 60 m.

    A) 15 m B) 60 m C) 80 m D) 120 m

    26. B

27. The ratio of $\frac{4}{3}$ to $\frac{3}{4}$ is $\frac{4}{3} \times \frac{4}{3} = \frac{16}{9}$.

    A) 1 **B)** $\frac{3}{4}$ C) $\frac{4}{3}$ D) $\frac{16}{9}$

    27. D

28. I bought an odd number of pens, so I bought an odd number of packs of 3. If I bought 1 pack of 3, I could have bought 2 packs of 8, 1 pack of 6, and 8 packs of 12. No other number of packs of 3 yields 12 packs.

    A) 1 B) 2 C) 3 D) 4

    28. B

29. $3^2 \times (2 \times 2 \times 2)^2 \times 5^2 = (3 \times 2)^2 \times 2^2 \times (5 \times 2)^2$.

    A) $\frac{1}{2}$ B) 2 C) 2^2 D) 2^3

    29. C

30. There is one "1" from 1 to 9, 11 "1"s from 10 to 19, one "1" in each of the next 8 groups of 10 integers, and one "1" in 100.

    A) 18 B) 19 C) 20 D) 21

    30. D

31. When expanded, $20^{10} = 10\,240\,000\,000\,000$. The difference between the product and the sum of the non-zero digits is $8 - 7 = 1$.

    A) 1 B) 2 C) 10^2 D) 2×10

    31. A

32. In the sequence 20, $\frac{19}{2}, \frac{18}{3}, \frac{17}{4}, \ldots$, each term after the first term is gotten by subtracting 1 from the previous term's numerator and adding 1 to the previous term's denominator. The only integers in this sequence are 20, 18/3, and 14/7.

    A) 1 B) 2 C) 3 D) 4

    32. C

33. The area of each rectangle is half of the area of the non-overlapping region plus the area of the square. Therefore, the area of each rectangle is $12/2 + 4 = 10$.

    A) 4 B) 6 C) 8 D) 10

    33. D

34. If the mean of three positive integers is 5, their sum is 15. The integers could be 5, 5, and 5.

    A) 105 B) 120 C) 125 D) 150

    34. C

35. Since the square root of 100 000 is between 316 and 317, 317 is the smallest such 3-digit integer.

    A) 5 B) 7 C) 9 D) 11

    35. D

The end of the contest **7**

Visit our Website at http://www.mathleague.com

SEVENTH GRADE MATHEMATICS CONTEST

Math League Press, P.O. Box 17, Tenafly, New Jersey 07670-0017

Information & Solutions

Tuesday, February 18 or 25, 2020

7

Contest Information

- **Solutions** Turn the page for detailed contest solutions (written in the question boxes) and letter answers (written in the *Answer Column* to the right of each question).

- **Scores** Please remember that *this is a contest, and not a test*—there is no "passing" or "failing" score. Few students score as high as 28 points (80% correct); students with half that, 14 points, *deserve commendation!*

- **Answers and Rating Scales** Turn to page 141 for the letter answers to each question and the rating scale for this contest.

1. $(2 \times 4 \times 8) \div 2 = 1 \times 2 \times 4 \times \underline{4}$.

   A) 1 B) 2 C) 4 D) 8

   Answer: 1. C

2. Sam spent $\frac{4}{5} \times \$80 = \frac{\$320}{5} = \$64$ on candy.

   A) \$16 B) \$32 C) \$64 D) \$300

   Answer: 2. C

3. The total value of 75 nickels is $75 \times 5¢ = 375¢$. The total number of quarters is $375¢ \div 25¢ = 15$.

   A) 3 B) 15 C) 25 D) 50

   Answer: 3. B

4. $100 \times 0.01 = (10 \times 10) \times 0.01 = 10 \times \underline{10 \times 0.01}$, and $10 \times 0.01 = \underline{0.1}$.

   A) 0 B) 0.01 C) 0.1 D) 1

   Answer: 4. C

5. Since $\frac{1}{4} \times \frac{3}{4} = \frac{3}{16}$ of the pets are dogs, the rest, $1 - \frac{1}{4} - \frac{3}{16} = \frac{9}{16}$, of Petunia's pets are birds.

   A) $\frac{9}{16}$ B) $\frac{8}{16}$ C) $\frac{7}{16}$ D) $\frac{6}{16}$

   Answer: 5. A

6. Round to the nearest 10 & 100, as shown. Choice D is 10 more when rounded to the nearest 10 than when rounded to the nearest 100.

   A) 100, 100 B) 100, 100 C) 100, 100 D) 110, 100

   Answer: 6. D

7. The reciprocal of one-fourth is 4, or <u>sixteen</u>-fourths.

   A) four B) eight C) ten D) sixteen

   Answer: 7. D

8. Hanni's Hal's handstand lasted $48 \times 60 =$ Hanni's lasted $2880 + 480 = 3360$ sec. $=$ $(3360 \div 60)$ minutes $= \underline{56}$ minutes.

   A) 56 B) 58 C) 86 D) 96

   Answer: 8. A

9. The 3 smallest consecutive integers that include 5 and an even integer are 3, 4, and 5. Their product is 60.

   A) 10 B) 30 C) 60 D) 720

   Answer: 9. C

10. If the perimeter is 12, the area is 9. All areas must be divisible by 9.

    A) 6 B) 8 C) 9 D) 12

    Answer: 10. C

11. If 2019 is the greatest of 2020 consecutive integers, then 0 must be the smallest. Any product with 0 as a factor equals 0.

    A) 0 B) negative C) prime D) odd

    Answer: 11. A

12. Subtracting 19 from a product of twenty 19s leaves nineteen 19s.

    A) 0 B) 20 C) 19^2 D) 20^2

    Answer: 12. C

13. Since the square root of 1000 is 31.6 . . . , $31^2 = 961$ is the greatest perfect square less than 1000. The sum of its digits is 16.

    A) 1 B) 4 C) 9 D) 16

    Answer: 13. D

Go on to the next page ⟹ **7**

14. The number of even numbers is $2020^2 \div 2 = (2020 \times 2020) \div 2 =$ 2020×1010.

A) 2020 B) 1010×1010 C) 1010×2020 D) 2020×2020

14. C

15. The average number of fish in my 4 fish tanks is 5. There are $4 \times 5 = 20$ fish in all. If 3 tanks have the minimum 1 fish each, the 4th has the maximum $20 - 3 = 17$ fish.

A) 17 B) 18 C) 19 D) 20

15. A

16. The least perimeter is when side-lengths are closest in value. Perimeter $= 4 \times 12 = 48$.

A) 48 B) 50 C) 52 D) 60

16. A

17. The ratio of \$6 books to \$8 books is 6:1. So $154 \div (6 + 1) = 154 \div 7 =$ 22 of the books were \$8 books. Ben spent $22 \times \$8 = \$176$ on \$8 books.

A) \$88 B) \$154 C) \$168 D) \$176

17. D

18. The least integer greater than 1 that has a positive remainder when divided by 2, 3, 4, 5, 6, 7, 8, or 9 is 11; $100 \div 11$ has a remainder of 1.

A) 1 B) 2 C) 3 D) 4

18. A

19. If I end 2020 at 100 cm tall, I will end 2021 at $100 + 0.1 \times 100 = 110$ cm tall, and 2022 at $110 + 0.1 \times 110 = 121$ cm, 21% taller than in 2020.

A) 11 B) 12 C) 20 D) 21

19. D

20. The 3 primes must include 2, the only even prime. The 3 consecutive primes that include 2 are 2, 3, and 5. Their sum is 10.

A) 5 B) 6 C) 10 D) 30

20. C

21. All of the choices equal $100^2 = 10000$, except D, which equals 10000^2.

A) $(10^2)^2$ B) $(10 \times 10)^2$ C) $10^2 \times 10^2$ D) $(10^2 \times 10^2)^2$

21. D

22. Of the integers 1 to 2020, $2020 \div 4 = 505$ are multiples of 4, and $2020 \div 5 = 404$ are multiples of 5.

A) 100 B) 101 C) 110 D) 111

22. B

23. From noon yesterday to 12 a.m. today $= 12 \times 60 =$ 720 min. Since $720 \div 7 = 102R6$, the faucet dripped 6 min. before 12 a.m., and again at 12:01 a.m.

A) 12:00 a.m. B) 12:01 a.m. C) 12:03 a.m. D) 12:07 a.m.

23. B

24. The square of the square of 3 is $3 \times 3 \times 3 \times 3$, with factors 1, 3, 9, 27, 81.

A) 1 B) 3 C) 4 D) 5

24. D

25. Simplify each choice as shown. The smallest positive value is C.

A) 0.1 B) 0.01 C) 0.001 D) 0.2

25. C

Go on to the next page ⟹

Solution	Answers
26. If Ky ran 400 m in 10 min., then Cy ran 1 km in 20 min. and 500 m in 10 min. The ratio of 400 m/10 min. to 500 m/10 min. is 4:5. A) 2:5 B) 1:2 C) 4:5 D) 4:1	26. C
27. Since $2019 \div 12 = 168R3$, each person picked a gumball 168 times. They stopped picking gumballs after the 12th person had picked their 168th gumball and there were 3 gumballs left. A) 1 B) 3 C) 7 D) 9	27. B
28. Multiples of 5 end in 5 or 0. No primes end in $5 + 1 = 6$. The primes less than 100 that end in $0 + 1 = 1$ are 11, 31, 41, 61, and 71. A) 5 B) 6 C) 7 D) 8	28. A
29. Increasing the sum of the ones digits by 2 changes the tens and hundreds digits. The original sum's ones digit must have been 8 or 9, and the original sum's tens digit must have been 9. Only choice C fits. A) 189 B) 197 C) 198 D) 209	29. C
30. The integer -9 is greater than -99 by 90, and $9 - 90 = -81$. A) -81 B) -90 C) -99 D) -107	30. A
31. The 1st 4 multiples: $4 \times 4 = 16$ digits. Multiples 5-49: $45 \times 5 = 225$ digits; 241 digits so far. The 240th digit is the tens digit of $49 \times 2020 = 98\,980$. A) 0 B) 2 C) 4 D) 8	31. D
32. There are $2020 \div 2 = 1010$ multiples of 2. Of the $2020 \div 5 = 404$ multiples of 5, $2020 \div (2 \times 5) = 202$ of them were already erased. There are $2020 - 1010 - (404 - 202) = 808$ numbers left. A) 606 B) 808 C) 1212 D) 1414	32. B
33. Choose 1s for the first 3 integers and 4s for the last 3: 1 1 1 **4** 4 4 4. A) 16 B) 19 C) 23 D) 28	33. B
34. $2^{2020} = (2^{1010})^2 = (2^{505})^4 = (2^{404})^5 = (2^{202})^{10} = (2^{101})^{20} = (2^{20})^{101} = (2^{10})^{202} = (2^5)^{404} = (2^4)^{505} = (2^2)^{1010} = (2^1)^{2020}$. There are 11 ways. A) 3 B) 6 C) 11 D) 1010	34. C
35. A $\frac{2}{3}$ rotation moves each car $\frac{2}{3} \times 24 = 16$ places clockwise. A $\frac{3}{4}$ rotation moves each $\frac{3}{4} \times 24 = 18$ places clockwise. Car #13 is 16 places behind car #5. Car #11 is 18 places behind car #5. A) 11, 12, 13 B) 13, 14, 15 C) 17, 18, 19 D) 21, 22, 23	35. A

The end of the contest 7

Visit our Website at http://www.mathleague.com

Math League Press, P.O. Box 17, Tenafly, New Jersey 07670-0017

Information & Solutions

Tuesday, February 16 or 23, 2021

7

Contest Information

- **Solutions** Turn the page for detailed contest solutions (written in the question boxes) and letter answers (written in the *Answer Column* to the right of each question).

- **Scores** Please remember that *this is a contest, and not a test*—there is no

2020-2021 7TH GRADE SOLUTIONS	*Answers*
1. 2000 + 200 + 20 + 2 = 2222 and 2222 − **201** = 2021. A) 1 B) 11 C) 21 D) 201	1. D
2. Each bag has 6 more baseballs than bats. If there are 18 more baseballs than bats, there are 3 bags. The number of balls is 3 × 9. A) 27 B) 36 C) 48 D) 54	2. A
3. 0.5 = 0.50 = 1/2 = 50%; 50/10 = 5. A) 0.50 B) $\frac{1}{2}$ C) $\frac{50}{10}$ D) 50%	3. C
4. 2020 m = 1 km + _?_ m = 1000 m + _?_ m = 1000 m + **1020** m. A) 1020 B) 2202 C) 2222 D) 2244	4. A
5. (1/2÷1/4) = (1/16÷1/32) = 2; so (1/2÷1/4)÷**1/8** = (1/16÷1/32)÷**1/8**. A) 1/2 B) 1/8 C) 1/32 D) 1/64	5. B
6. In each classroom in my school there are 3 × 6 = 18 chairs. If there are 270 chairs in all, there are 270 ÷ 18 = 15 classrooms. A) 15 B) 30 C) 45 D) 90	6. A
7. There are 11 such numbers: 9 × 12, 9 × 13, 9 × 14, . . . , 9 × 21, 9 × 22. A) 9 B) 10 C) 11 D) 12	7. C
8. 2021 ÷ 1000 = 2.021; rounded to the nearest tenth, it is 2.0. A) 2.0 B) 2.02 C) 20.2 D) 20.21	8. A
9. Since 2222 = 2 × 1111, 2222 has only 1 factor of 2. A) 2000 B) 2200 C) 2220 D) 2222	9. D
10. The sum of the leg-lengths must be an even number > 10. If the sum is 12, perimeter is 22. A) 19 B) 20 C) 21 D) 22	10. D
11. Pat correctly answers 2/5 of the first 10 questions, which is 4 questions. There are 25 remaining, for a total of 25 + 4 = 29 correct. A) 14 B) 25 C) 29 D) 31	11. C
12. After 6 minutes, Emma counted to 360. After another 2 minutes, she counted down 120 numbers from 360, so she ended on 240. A) 120 B) 240 C) 360 D) 480	12. B
13. Write each to the thousandths place; the choice closest to 0 is choice C. A) 0.100 B) 0.090 C) 0.010 D) 0.062	13. C

Go on to the next page ⟫➡ **7**

	Answers
14. Beginning at 5:00 AM, my sprinkler is off during the following time ranges: 5:20-5:25, 5:45-5:50, 6:10-6:15, 6:35-6:40, 7-7:05, 7:25-7:30, 7:50-7:55, 8:15-8:20, 8:40-8:45, 9:05-9:10, 9:30-9:35, 9:55-10, 10:20-10:25, 10:45-10:50, and 11:10-11:15. Only 9:06 AM falls within these ranges. A) 8:06 AM B) 9:06 AM C) 10:06 AM D) 11:06 AM	14. B
15. I practice catching footballs for 72 minutes every day. That is 72/1440 = 1/20 of each day. A) 1/3 B) 5/144 C) 1/18 D) 1/20	15. D
16. Divide each choice by $5^{20} \div 6^{20}$ and find least quotient. A) 1 B) 5/6 C) 25/216 D) 125/36	16. C
17. Since 12 minutes = 720 seconds, it will take 720 days. 720 days is about 10 days short of 2 years, so it will be December. A) January B) June C) July D) December	17. D
18. Since 0.125 = 1/8, we must multiply by a multiple of 8. A) 5555 B) 6666 C) 7777 D) 8888	18. D
19. $2020 = 2\times2\times5\times101$; 2020 is divisible by 3 different prime numbers. A) 2 B) 3 C) 4 D) 5	19. B
20. Sally Speedster read 80 + 20 = 100 pages of a book in 40 + 20 = 60 minutes. She read at an average rate of 100 pages/60 min = 100/60 pages/min = 5/3 pages/min. A) 3/2 B) 8/5 C) 5/3 D) 7/4	20. C
21. Find the smallest whole number that is divisible by 7 and is 1 more than a multiple of 10. The smallest such number is 21. A) 0 and 25 B) 26 and 50 C) 51 and 75 D) 76 and 100	21. A
22. 50% of 50% = 1/2 of 1/2 = 1/4 = one quarter of 1. A) $\frac{1}{4}$ B) $\frac{1}{2}$ C) 1 D) 25%	22. C
23. The number 72 is divisible by 1, 2, 3, 4, 6, 8, 9, 12, 18, 24, 36, and 72. A) 49 B) 64 C) 72 D) 81	23. C
24. $\left(\frac{9}{4}\right)^2 + 3^2 = \frac{81}{16} + 9 = \frac{225}{16} = \left(\frac{15}{4}\right)^2$. A) $\left(\frac{15}{4}\right)^2$ B) 4^2 C) 5^2 D) $\left(\frac{21}{4}\right)^2$	24. A
25. 1547 beach balls were distributed among all the area beaches. Since 1547 ÷ 91 is a whole number, there could be 91 beaches. A) 21 beaches B) 35 beaches C) 57 beaches D) 91 beaches	25. D

Go on to the next page ⟹ 7

Solution	Answers
26. I have exactly 15 coins, each of which is a nickel, dime, or quarter. If I have exactly 2 dollars, I could have 2 nickels, 9 dimes, and 4 quarters. A) 4 B) 5 C) 6 D) 7	26. A
27. Mo and Jo danced together for 30 seconds. Since 50 minutes = 3000 seconds, Mo and Jo danced together $30/3000 \times 100 = 1$ percent of the time. A) 1 B) 5 C) 6 D) 10	27. A
28. $1 \div 7 = 0.\overline{142857}$; $2021 \div 6 = 336R5$. The 5th digit of 142857 is 5. A) 1 B) 2 C) 4 D) 5	28. D
29. I bowled four games and averaged 130 points per game, so my total for the 4 games was 520. Since the average of my 3 highest games was 140, their total was 420. So my lowest score was 100. My other scores could have been 101, 102, and $420 - (101 + 102) = 217$. A) 150 B) 169 C) 197 D) 217	29. D
30. The reciprocals of 3 and 4 are 1/3 and 1/4. Their sum is 7/12. A) 1/7 B) 1/5 C) 7/12 D) 12/5	30. C
31. The sum of the measures of adjacent angles of a rhombus is 180°. The measures of these angles are 60° and 120° in this case. A) 45° B) 60° C) 72° D) 120°	31. B
32. Number rungs from 1 to 16. 1st minute: Zoe goes to rung 11 then back to 15; Dafne goes to 4 and then back to 2. 2nd minute: Zoe to 10 and back to 14; Dafne to 5 and back to 3. Continuing, during the 5th minute, Zoe and Dafne both reach rungs 7 and 8. A) 2nd B) 5th C) 7th D) 8th	32. B
33. 1 min.= 6°. At 1:10, the min. hand is 60° past 12 and the hr. hand is $30° + 1/6(30°)$ past 12. The larger angle is $360° - (60° - 35°) = 335°$. A) 300° B) 315° C) 330° D) 335°	33. D
34. The only prime numbers whose reciprocals are terminating decimals are 2 and 5. A) 1 B) 2 C) 3 D) 4	34. B
35. Writing last 2 digits only: $(2^{10})^8 = (24)^8 = (76)^4 = 76$. The sum is 13. A) 9 B) 11 C) 13 D) 15	35. C

The end of the contest 7

Visit our Web site at http://www.mathleague.com

8th Grade Solutions

2016-2017 through 2020-2021

Information & Solutions

8

Tuesday, February 21 or 28, 2017

Contest Information

- **Solutions** Turn the page for detailed contest solutions (written in the question boxes) and letter answers (written in the *Answer Column* to the right of each question).

- **Scores** Please remember that *this is a contest, and not a test*—there is no "passing" or "failing" score. Few students score as high as 28 points (80% correct); students with half that, 14 points, *deserve commendation!*

- **Answers and Rating Scales** Turn to page 143 for the letter answers to each question and the rating scale for this contest.

Solution	Answers
1. $(2\times22\times222)\div(1\times11\times111) = (2\div1)\times(22\div11)\times(222\div111) = 2\times2\times2$. A) 2 B) 2 + 2 + 2 C) 2 × 2 × 2 D) 1 + 11 + 111	1. C
2. $4 + (4 \div 4) \times 4 - 4 = 4 + (1 \times 4) - 4 = 4 + 4 - 4$. A) 0 B) 4 C) 8 D) 16	2. B
3. There are 500 safe deposit boxes at my bank. Of these 500 boxes, 10 are multiples of 50. The percent that are multiples of 50 is 2%. A) 0.1% B) 0.2% C) 2% D) 10%	3. C
4. One-tenth of 1 = 0.1, which is 1/10 more than 0. A) -0.1 B) 0 C) 0.1 D) 1	4. B
5. The reciprocal of 11/10 is 10/11 ≈ 0.91. Choice D is closest. A) 0.11 B) 0.1 C) 1.1 D) 1.0	5. D
6. The only 3-digit integers that can be written as the product of three consecutive primes are $3\times5\times7 = 105$ and $5\times7\times11 = 385$. A) 0 B) 1 C) 2 D) 3	6. C
7. The least common multiple of 16 and 24 is 48; 4 + 8 = 12. A) 5 B) 8 C) 12 D) 16	7. C
8. All factors of 2017^{2017} are odd; they are 1 and 2017^1 to 2017^{2017}. A) 0 B) 1 C) 2017 D) 2018	8. D
9. The side-length of a square with area 64 cm^2 is 8 cm. The side-length of a cube with volume 64 cm^3 is 4 cm. The ratio is 2:1. A) 2:3 B) 1:1 C) 2:1 D) 4:1	9. C
10. Consecutive angles of a rhombus are supplementary. The angles could have measures of 85°, 95°, 85°, and 95°. A) rhombus B) square C) triangle D) pentagon	10. A
11. Since $0.1\div(0.01 + 0.001) = 0.1\div 0.011 = 100\div11$, the denominator of $\frac{1}{10} \div \left(\frac{1}{100}+\frac{1}{1000}\right)$ is 11. A) 10 B) 11 C) 1000 D) 1100	11. B
12. The first product must be the product of the first 2017 positive integers and the second product must be the same integers except 1. The sum of the greatest integers in each group is 2017 + 2017 = 4034. A) 4032 B) 4033 C) 4034 D) 4035	12. C

Go on to the next page ➡ **8**

13. Consider the 4 vertical edges; each one is parallel to the other 3.

A) 1 B) 2 C) 3 D) 4

13. C

14. The sides of the rectangle could have lengths 3 and 4. By the Pythagorean Th., a diagonal would have length 5.

A) 5 B) 8 C) 12 D) 13

14. A

15. If 30% are quarters, the value is divisible by 75¢+70¢ = $1.45. Choice C is divisible by $1.45.

A) $10.25 B) $11.80 C) $13.05 D) $14.75

15. C

16. 1.1 is 0.1 more than 1—that's the closest one.

A) 1.1 B) 1.1^{10} C) 0.011 D) 1.11

16. A

17. The number of chairs in each row equals the number of rows, so the product of these 2 numbers is a perfect square. The largest perfect square that is less than 999 is 961; $999 - 961 = 38$.

A) 31 B) 33 C) 36 D) 38

17. D

18. Let's use 2, 3, and 5. Then 1, 6, 10, 15, and 30 are not prime.

A) 2 B) 3 C) 4 D) 5

18. D

19. The 500th group he typed ends with 4000. The numbers are 3993, 3994, 3995, . . . , 3999, and 4000. Their sum is 31 972.

A) 30 864 B) 31 972 C) 32 144 D) 34 032

19. B

20. If the lengths were each 1 originally, the new volume would be 27.

A) 3 B) 2^3 C) 3^2 D) 3^3

20. D

21. The only factors of 51^6 that are prime are 3 and 17. (1 is not prime.)

A) 0 B) 1 C) 2 D) 3

21. C

22. In a group of 10 students, each student shakes hands with 9 other students. That's 90 handshakes, but each one has been counted twice.

A) 19 B) 20 C) 45 D) 90

22. C

23. A side of the square has length is 4; its area is 16. The area of the circle is 4π. Half the circle is inside the square; area outside is $16 - 2\pi$.

A) 2.28 B) 3.34 C) 3.43 D) 9.72

23. D

24. The sum of Zak's 6 scores was $90 \times 6 = 540$. The sum of his 1st 5 scores was 440, for an average of $440 \div 5 = 88$.

A) 80 B) 85 C) 88 D) 92

24. C

25. A fraction with denominator 0 is undefined.

A) 0 B) 0.001 C) 1 D) 1 000 000

25. A

Go on to the next page ⟹ 8

26. The 8 factors of 40 are 1, 2, 4, 5, 8, 10, 20, and 40; 2 are odd.

    A) $\frac{1}{5}$ B) $\frac{1}{4}$ C) $\frac{1}{3}$ D) $\frac{1}{2}$

    26. B

27. Ted is 13 years old since $13+12\times13=169$. Age in years + age in months = 13×age in years. For next occurrence, multiply age in years by 4, the smallest perfect square > 1: 13×13×4. In years his age will be 13×4 = 52, 39 years later.

    A) 3 B) 13 C) 26 D) 39

    27. D

28. The chance of rain each day for seven days after Tuesday is 50%, 25%, 12.5%, 6.25%, 3.125%, 1.5625%, 0.78125%. Seven days after Tuesday is another Tuesday.

    A) Sunday B) Monday C) Tuesday D) Wednesday

    28. C

29. The numbers have 2 or 4 in the 1's place and any of the 5 digits in the other 3 places. That's $5\times5\times5\times2=250$ possibilities.

    A) 64 B) 125 C) 128 D) 250

    29. D

30. The ratio of nuts to raisins changed from 10:6 to 9:6 after removing 50 nuts. Thus 50 was 1/10 of the number of nuts. There were 500 nuts and 300 raisins to start. I now have 450 nuts, which is 150 more than 300.

    A) 120 B) 150 C) 160 D) 180

    30. B

31. The volume of water in the aquarium is the product of the area of its base and its depth. Since the area of the base is 16 m^2, its depth must be 1/16 m.

    A) $\frac{1}{8}$ m B) $\frac{1}{4}$ m C) $\frac{1}{4\times4}$ m D) $\frac{1}{4\times4\times8}$ m

    31. C

32. Since $2^{2017}\times5^{2007}=2^{10}\times(2^{2007}\times5^{2007})=1024\times10^{2007}$, sum of non-zero digits is 1+2+4 = 7.

    A) 7 B) 8 C) 9 D) 10

    32. A

33. There are 26×25×24×23 passwords with 4 different letters. There are 4×3×2×1 passwords with M, A, T, H. The ratio is 1/(26×25×23).

    A) $\frac{1}{23\times24\times25}$ B) $\frac{1}{23\times25\times26}$ C) $\frac{1}{23\times24\times26}$ D) $\frac{1}{24\times25\times26}$

    33. B

34. Of every 180 votes mailed in, Abe receive 80 and George received 100. If 180 is 45%, then 400 is 100%. Of every 220 votes remaining, Abe must get 120. Abe must get 120/220 = 6/11 of the remaining votes.

    A) $\frac{6}{11}$ B) $\frac{3}{5}$ C) $\frac{2}{3}$ D) $\frac{3}{4}$

    34. A

35. The 1st 9 integers use 9 digits and the next 90 use 180. There are 1828 more digits. Since $1828\div3=609$ R1, the 2017th digit is the 1st digit of 709.

    A) 4 B) 5 C) 6 D) 7

    35. D

The end of the contest 8

Visit our Website at http://www.mathleague.com

EIGHTH GRADE MATHEMATICS CONTEST

Math League Press, P.O. Box 17, Tenafly, New Jersey 07670-0017

Information & Solutions

8

Tuesday, February 20 or 27, 2017

Contest Information

- **Solutions** Turn the page for detailed contest solutions (written in the question boxes) and letter answers (written in the *Answer Column* to the right of each question).

- **Scores** Please remember that *this is a contest, and not a test*—there is no "passing" or "failing" score. Few students score as high as 28 points (80% correct); students with half that, 14 points, *deserve commendation!*

- **Answers and Rating Scales** Turn to page 144 for the letter answers to each question and the rating scale for this contest.

2017-2018 8TH GRADE CONTEST SOLUTIONS	***Answers***
1. $5 + 5 - 5 + (5 \times 5) + (5 \div 5) + 5 = 5 + 25 + 1 + 5 = 36$. A) 0 B) 1 C) 25 D) 36	1. D
2. The product of 8 and 12 and 4 is 384. A) 24 B) 384 C) 768 D) 1804	2. B
3. The only such integers are 150, 300, 450, 600, 750, and 900. There are 6 of them. A) 1 B) 2 C) 3 D) 6	3. D
4. If one runs 2 m/sec. and the other runs 4 m/sec., the distance between them decreases 6 m/sec. In order for the distance to decrease by 18 m, they must run for 3 seconds. A) 2 B) 3 C) 4 D) 8	4. B
5. The diagonals form 4 large $\triangle$s, each 1/2 the rectangle, and 4 small $\triangle$s. A) 2 B) 4 C) 6 D) 8	5. D
6. $(1 + 2 + 3 + \ldots + 2016 + 2017) \div 2017 = (1 + 2017) \div 2 = 1009$. A) 1008 B) 1009 C) 2017 D) 2018	6. B
7. The only prime factors of 72 are 2 and 3; their sum is 5. A) 2 B) 3 C) 5 D) 6	7. C
8. 100% of a number + 20% of that number is 120% of that number. A) 20% B) 80% C) 120% D) 200%	8. C
9. $100 in nickels is 2000 coins; $100 in dimes is 1000 coins. A) 100 B) 200 C) 1000 D) 2000	9. C
10. The difference between the largest and smallest integer is 2017. A) 1009 B) 2017 C) 2018 D) 2019	10. B
11. Since $\frac{123456789}{100} = 1234567.89$, it has 2 non-zero digits to the right of the decimal. A) 2 B) 3 C) 6 D) 7	11. A
12. Only 2 songs were sung by the entire choir, so 22 songs were sung by individual choir members. The choir must have 22 members. A) 8 B) 11 C) 12 D) 22	12. D
13. The product of 2017 and 2018 is a multiple of 2018. Since it is divisible by 2018, the remainder when it is divided by 2018 is 0. A) 0 B) 1 C) 2017 D) 2018	13. A

Go on to the next page ➠ **8**

14. My armful of gumballs weighs 4% less since I dropped one gumball. Since 4% = 4/100 = 1/25, I have 25 − 1 = 24 gumballs in my arms.

A) 23 B) 24 C) 25 D) 26

14. B

15. The least 2-digit integer that is both a perfect square and a perfect cube is 64; 6 + 4 = 10.

A) 7 B) 8 C) 9 D) 10

15. D

16. The degree-measures of the 2 complementary angles add up to 90, leaving 90 for the 3rd one.

A) equilateral B) right C) scalene D) obtuse

16. B

17. Since 1 + 0.1 = 1.1, it is the sum with the greatest value.

A) $1 + 0.1$ B) $1^{10} + 0.1^{10}$ C) $1^{100} + 0.1^{100}$ D) $1^{1000} + 0.1^{1000}$

17. A

18. The sum of the lengths of all 12 edges of the cube is 144 cm. Each edge is 12 cm, and the area of one face is 12^2 cm^2 = 144 cm^2.

A) 144 cm^2 B) 196 cm^2 C) 256 cm^2 D) 324 cm^2

18. A

19. Since 815 min. = 13 hr. 35 min., the time is 9:50 A.M.

A) 3:15 A.M. B) 9:50 A.M. C) 3:15 P.M. D) 9:50 P.M.

19. B

20. The number 180 has 18 divisors, and the number 120 has 16 divisors.

A) 0 B) 2 C) 30 D) 60

20. B

21. The 8 houses have consecutive integer addresses that add up to 1500. The average is 187.5, so the address with the greatest value is 191.

A) 184 B) 187 C) 188 D) 191

21. D

22. $\frac{10}{3} = \frac{3}{1} + \frac{1}{3}$.

A) $\frac{7}{3}$ B) $\frac{8}{3}$ C) $\frac{9}{3}$ D) $\frac{10}{3}$

22. D

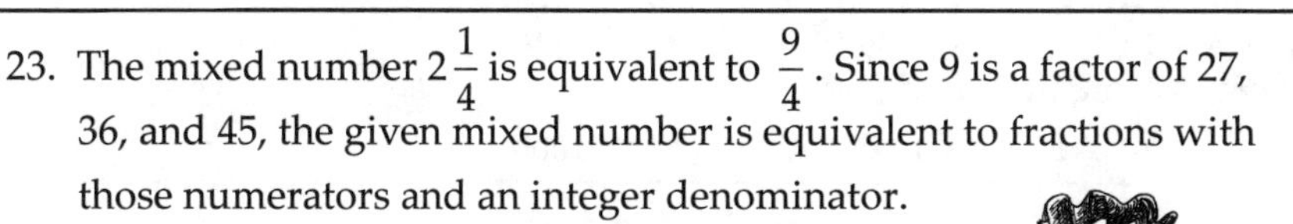

23. The mixed number $2\frac{1}{4}$ is equivalent to $\frac{9}{4}$. Since 9 is a factor of 27, 36, and 45, the given mixed number is equivalent to fractions with those numerators and an integer denominator.

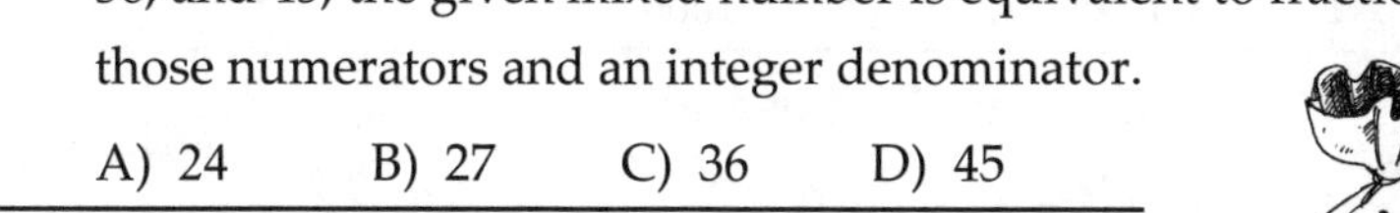

A) 24 B) 27 C) 36 D) 45

23. A

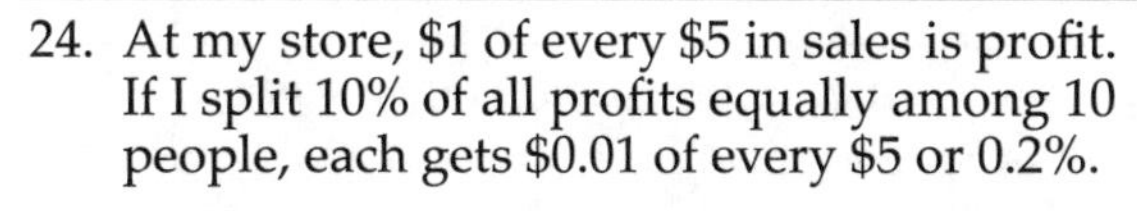

24. At my store, \$1 of every \$5 in sales is profit. If I split 10% of all profits equally among 10 people, each gets \$0.01 of every \$5 or 0.2%.

A) 0.2 B) 2 C) 5 D) 20

24. A

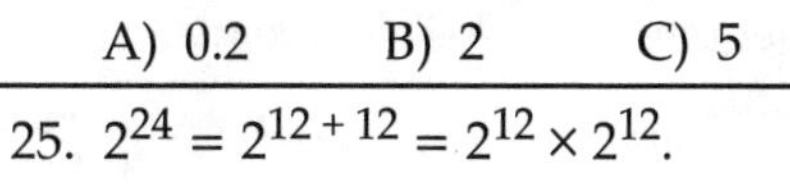

25. $2^{24} = 2^{12+12} = 2^{12} \times 2^{12}$.

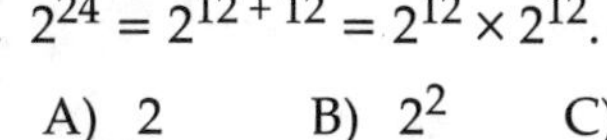

A) 2 B) 2^2 C) 2^{12} D) 2^{36}

25. C

Go on to the next page ⟹ **8**

	Answers
26. It's D since D has the smallest numerator and the largest denominator. A) $\frac{3^{100}}{4}$ B) $\left(\frac{3}{4}\right)^{100}$ C) $\frac{3}{4}$ D) $\frac{3}{4^{100}}$	26. D
27. Of the 99 positive integers less than 100, only 30, 42, 66, 70, and 78 are the product of exactly 3 different prime numbers. A) $\frac{1}{99}$ B) $\frac{4}{99}$ C) $\frac{5}{99}$ D) $\frac{8}{99}$	27. C
28. The least and greatest numbers must be even. Only choice A is the sum of two even numbers that differ by 2. (The ticket numbers are 8, 9, 10.) A) 18 B) 20 C) 24 D) 28	28. A
29. Eve counted to $4^{60} = (2^2)^{60} = 2^{120}$ by consecutive powers of 2, counting $2^1, 2^2, 2^3, \ldots, 2^{120}$. That's 120 powers of 2. A) 30 B) 120 C) 240 D) 3600	29. B
30. The ones digit must be a 2, and there are 4 choices for the other digits. There are 1, 4, 16, 64, 256, 1024 numbers with 1, 2, 3, 4, 5, 6 digits. A) 1365 B) 3906 C) 5400 D) 19530	30. A
31. If 6 identical machines can fill 80 bottles of soda in 12 seconds, 36 machines can fill 80 bottles in 2 seconds. It would take 36 machines 3 times as long to fill 240 bottles of soda. A) 6 B) 12 C) 18 D) 24	31. A
32. $100\% - 76\% = 24\%$ of my favorite songs were released in 2017 or later. Since 42% of my favorite songs were released in 2016 or later, $42\% - 24\% = 18\%$ of my favorite songs were released in 2016 alone. A) 18% B) 24% C) 34% D) 58%	32. A
33. The positive integers less than 10^6 that are perfect squares are $1^2, 2^2, 3^2, \ldots, 998^2, 999^2$. Of these 999 numbers, 499 are even and 500 are odd. A) 1:1 B) 2:1 C) 499:500 D) 999:1000	33. C
34. $(4\times504 + 1)^{2018}$ is 1 more than a multiple of 4; add 3 to get a multiple. A) $2017^{2018}+1$ B) $2017^{2018}+3$ C) $2017^{2018}+5$ D) $2018^{2017}+1$	34. B
35. If the sum of the degree-measures of 2 angles of a parallelogram is 108, the degree-measure of each is 54, and the others are each 126. A) 72 degrees B) 162 degrees C) 234 degrees D) 252 degrees	35. C

The end of the contest  **8**

Visit our Website at http://www.mathleague.com

EIGHTH GRADE MATHEMATICS CONTEST

Math League Press, P.O. Box 17, Tenafly, New Jersey 07670-0017

Information & Solutions

8

Tuesday, February 19 or 26, 2019

Contest Information

- **Solutions** Turn the page for detailed contest solutions (written in the question boxes) and letter answers (written in the *Answer Column* to the right of each question).

- **Scores** Please remember that *this is a contest, and not a test*—there is no "passing" or "failing" score. Few students score as high as 28 points (80% correct); students with half that, 14 points, *deserve commendation!*

- **Answers and Rating Scales** Turn to page 144 for the letter answers

1. $(4 \times 6 \times 8 \times 10) \div (6 \times 8 \times 10) = 4 \times 1 \times 1 \times 1 = 4.$

   A) 3 B) 4 C) 12 D) $3\times6\times8\times10$

   1. B

2. $2 \div 3 = 0.666\ldots$; this rounds to 0.67.

   A) 0.33 B) 0.66 C) 0.67 D) 0.70

   2. C

3. Their ages in days are consecutive integers. Since $132 = 11 \times 12$, the product of their ages in days could be 132.

   A) 33 B) 132 C) 245 D) 246

   3. B

4. The largest even divisor of 200 is 200, and the largest odd divisor of 200 is 25; $200 \div 25 = 8$.

   A) 4 B) 8 C) 20 D) 200

   4. B

5. An equilateral triangle with integer side-lengths has a perimeter that is a multiple of 3. The area of the square must also be a multiple of 3. If the length of a side of the square is 12, its area is 144.

   A) 12 B) 10 C) 8 D) 4

   5. A

6. We can pay \$12.50 using 50 quarters. That leaves \$0.10, which I can pay using one dime. The smallest number of coins is 51.

   A) 51 B) 52 C) 54 D) 55

   6. A

7. Since the sum of the digits of 2019 is divisible by 3, 2019 is also.

   A) 1 B) 3 C) 19 D) 673

   7. B

8. Since it is possible that the four integers do not include a multiple of 5, their product might not be divisible by a multiple of 5.

   A) 4 B) 6 C) 10 D) 12

   8. C

9. There are 28 days in 4 weeks. There are 24×28 hours in 28 days.

   A) 48 B) 96 C) 336 D) 672

   9. D

10. Try each choice and find the correct one. Since 10 divided by 1/10 is 100, choice D is correct.

    A) $\frac{1}{10}$ B) $\frac{1}{5}$ C) $\frac{1}{2}$ D) 10

    10. D

11. The height of the smoke is 100000 cm. To convert to km, divide by $10^2 \times 10^3 = 10^5$.

    A) 1 B) 10 C) 100 D) 1000

    11. A

12. Since $180° \div (4+5+6) = 180° \div (15) = 12°$, the measures are $4\times12° = 48°$, $5\times12° = 60°$, and $6\times12° = 72°$. Finally, $72° - 48° = 24°$.

    A) 12° B) 24° C) 30° D) 36°

    12. B

Go on to the next page ⟹ **8**

	Answers
13. The population of a town started at 1000, then went up to 1100, then down to 880, then up to 968. A) 968 B) 972 C) 1000 D) 1024	13. A
14. Divide each choice by 5. The quotients are 10, 15, 20, and 25. Since 15 + 60 is 75, choice B is correct. A) 50 B) 75 C) 100 D) 125	14. B
15. Each pair of angles in any rectangle is supplementary. A) triangle B) square C) rectangle D) hexagon	15. C
16. Drop the zeroes and evaluate: choices become 64, 243, 256, and 125. A) 2^{600} B) 3^{500} C) 4^{400} D) 5^{300}	16. A
17. $(2^{100} \times 4^{50}) \div 2 = (2^{100} \times 2^{100}) \div 2 = 2^{200} \div 2^1 = 2^{199}$. A) 2^{75} B) 2^{100} C) 2^{149} D) 2^{199}	17. D
18. The pattern for the ones digits of powers of 3 is 39713971..., and the 333rd digit is 3. A) 1 B) 3 C) 7 D) 9	18. B
19. On a series of tests, Gus got 100 once, 90 twice, and 80 five times. The total of these 8 tests is 680, and the average is 85. A) 80 B) 85 C) 90 D) 92	19. B
20. The product of 1 and 5 is 5. A) 5 B) 10 C) 35 D) 40	20. A
21. The probability of heads then tails then heads is $0.5 \times 0.5 \times 0.5 = 0.125$. A) 0.125 B) 0.25 C) 0.375 D) 0.5	21. A
22. There are 45 days from January 31 through March 16. Rui ate 180 jellybeans in those 45 days, or 4 jellybeans each day. There are 30 days from January 1 through January 30. Rui ate 120 jellybeans on those days, so Rui had 560 + 120 jellybeans on January 1. A) 600 B) 650 C) 680 D) 740	22. C
23. Jake used 40 boxes of tissues a day or 5760 tissues. Since $5760 \div 24 = 240$, he used 240 per hour or 4 per minute. A) 2 B) 3 C) 4 D) 5	23. C
24. $5184 = 64 \times 81$; its odd divisors are 1, 3, 9, 27, and 81. A) 1 B) 2 C) 4 D) 5	24. D
25. Only choice A is an even multiple of 5. A) 120 B) 125 C) 164 D) 212	25. A

Go on to the next page ➠ **8**

26. From 1 to 9 is one 1; from 10 to 19 is 11 1s; from 20 to 99 is 8 1s; from 100 to 109 is 11 1s, and from 110 to 111 is 5 1s. All together, we have (1 + 11 + 8 + 11 + 5) 1s. That is a total of 36 1s.

A) 12 B) 22 C) 24 D) 36

26. D

27. The whole numbers with squares between 2 and 200 are 2, 3, 4, 5, . . . , 13, and 14. There are 13.

A) 12 B) 13 C) 24 D) 26

27. B

28. A baker is cutting circular cookies out of a flat rectangle of cookie dough. If the rectangle is 200 cm by 100 cm and the cookies have diameter 20 cm, the baker can cut 10 rows, with 5 cookies in each row.

A) 50 B) 63 C) 64 D) 200

28. A

29. 0.02% of 20% = 0.00004; 200% of 2000 = 4000 = $0.00004 \times 100\,000\,000$.

A) 1000 B) 100 000 C) 1 000 000 D) 100 000 000

29. D

30. Since 3% of 1200 kg plus 6% of 2400 kg is 180 kg, and 40% of 100 kg is 40 kg, the remaining 3500 kg of ore has 140 kg of gold. Since 140 divided by 3500 = 0.04, the remaining ore will be 4% gold.

A) 2% B) 3% C) 4% D) 5%

30. C

31. There are 12 face diagonals and 4 diagonals passing through the interior.

A) 12 B) 14 C) 16 D) 24

31. C

32. Pick the hundreds digit, then the ones digit, then the tens digit. Based on the hundreds digit being even or odd, the count is $3\times4\times7 + 2\times5\times7$.

A) 154 B) 175 C) 185 D) 200

32. A

33. The whole-number factors of 36 are 1 and 36, 2 and 18, 3 and 12, 4 and 9, and 6. The product of their squares is 36^9.

A) 36^2 B) 36^4 C) 36^8 D) 36^9

33. D

34. When the four members of the Beaverton family carry a log, each has a probability of **not** tripping of 0.98, The probability of none of them tripping is $0.98 \times 0.98 \times 0.98 \times 0.98 = (0.98)^4$.

A) $1-(0.02)^4$ B) $(0.98)^4$ C) $(0.02)^4$ D) $1-(0.98)^4$

34. B

35. The largest prime factor of the product of all even numbers from 2 to 200 is the largest prime less than $200 \div 2 = 100$, which is 97.

A) 47 B) 97 C) 199 D) 2019

35. B

The end of the contest **8**

Visit our Website at http://www.mathleague.com

EIGHTH GRADE MATHEMATICS CONTEST

Math League Press, P.O. Box 17, Tenafly, New Jersey 07670-0017

Information & Solutions

8

Tuesday, February 18 or 25, 2020

Contest Information

- **Solutions** Turn the page for detailed contest solutions (written in the question boxes) and letter answers (written in the *Answer Column* to the right of each question).

- **Scores** Please remember that *this is a contest, and not a test*—there is no "passing" or "failing" score. Few students score as high as 28 points (80% correct); students with half that, 14 points, *deserve commendation!*

- **Answers and Rating Scales** Turn to page 146 for the letter answers to each question and the rating scale for this contest.

1. $8\,000\,000 \times 16\,000\,000 = \underline{8 \times 16} \times 1\,000\,000\,000\,000$; ___?___ $= 8 \times 16 = 128$.

   A) 24 B) 32 C) 64 D) 128

   1. D

2. Triplet sisters are celebrating their birthday today! Since they are all the same age, the sum of their three ages must be divisible by 3.

   A) 6 B) 13 C) 17 D) 80

   2. A

3. The number of fish in a school is equal to the cube of an integer. There could be $\underline{3 \times 3 \times 3} = 27$ fish in the school.

   A) 16 B) 27 C) 36 D) 101

   3. B

4. 10 m/sec. = 600 m/min. = 36000 m/hr. = 36 km/hr.

   A) 10 B) 36 C) 60 D) 72

   4. B

5. $0.3/100 \times 30/100 \times 30\,000 = 270\,000/10\,000 = 27$.

   A) 27 B) 270 C) 2700 D) 27 000

   5. A

6. $2020 = 2 \times 2 \times 5 \times 101$. Of these prime factors, 101 is greatest.

   A) 2 B) 5 C) 101 D) 202

   6. C

7. Match up the numbers in the two groups. Each group contains 2 through 2019, so the only difference is $2020 - 1 = 2019$.

   A) 1 B) 2 C) 2019 D) 2020

   7. C

8. $1/6 \div 1/3 = 1/6 \times 3/1 = 3/6 = 1/2$.

   A) $\frac{1}{18}$ B) $\frac{1}{2}$ C) 2 D) 18

   8. B

9. $(2^{2019} \times 2^{2020})^2 = (2^{4039})^2 = 2^{8078}$.

   A) 2^{4041} B) 2^{6059} C) 2^{8078} D) $2^{4078382}$

   9. C

10. $36 = (2^2 \times 3^2)$ and $75 = (3 \times 5^2)$, so their l.c.m is $2^2 \times 3^2 \times 5^2 = 900$, which has square root 30.

    A) 30 B) 50 C) 52 D) 2700

    10. A

11. For every 5 whole apples there are 2 bitten. So there could be 10 and 4, or 15 and 6, etc. Since 15 and 6 differ by 9, 6 apples are bitten.

    A) 3 B) 6 C) 15 D) 18

    11. B

12. The maximum area is that of a square with sides of length $24/4 = 6$.

    A) 20 B) 24 C) 36 D) 144

    12. C

13. $6 < 10000 \div (60 \times 24) < 7$, so it will be 7 days before Jan. 1.

    A) Dec. 27 B) Dec. 26 C) Dec. 25 D) Dec. 24

    13. C

Go on to the next page ⟹ **8**

Solution	Answers
14. The 20 human clients averaging 80 kg each total 1600 kg, and the 30 porcupine clients averaging 10 kg each total 300 kg. The overall average is $(1600 + 300) \div 50 = 38$. A) 38 kg B) 42 kg C) 45 kg D) 52 kg	14. A
15. $-2^4 + 2^4 \times (-2)^4 + 2^4 = -16 + 16 \times 16 + 16 = 256$. A) 16 B) 256 C) 288 D) 528	15. B
16. The sum of 3 odd numbers is odd; dividing an odd by 2 leaves R1. A) 0 B) 1 C) 2 D) 3	16. B
17. The hundredths digit of 8765.4321 is 3 and the thousands digit is 8, so the difference is $8 - 3 = 5$. A) 1 B) 2 C) 4 D) 5	17. D
18. If \$60 is 120% of the online price, then \$10 is 20% of it, so the online price is \$50. When shipping fees of 10% are added, the online cost increases by \$5. I spent \$60 – \$55 = \$5 more than the online cost. A) \$5.00 B) \$6.00 C) \$7.20 D) \$10.00	18. A
19. The ones digit of an even power of 19 is always a 1. A) 9 B) 7 C) 3 D) 1	19. D
20. Use an example. A 4-by-5 rectangle has perimeter $4+5+4+5 = 18$. Tripling each side gives perimeter $12+15+12+15 = 54$. $54 \div 18 = 3$. A) 3 B) 6 C) 9 D) 12	20. A
21. Paola's age has an odd number of divisors, so it is a perfect square. A prime squared has only 3 divisors, but Paola's age has 5 divisors. A) a prime squared B) a non-prime squared C) an odd number D) an even number	21. B
22. abc, abd, abe, abf, acd, ace, acf, ade, adf, aef, bcd, bce, bcf, bde, bdf, bef, cde, cdf, cef, def. A) 18 B) 20 C) 40 D) 120	22. B
23. Each number is 4 less than the previous multiple of 4. The 2020th number is $8100 - 2019 \times 4 = 24$. A) 4 B) 20 C) 24 D) 64	23. C
24. To play 250% more than 200 hours, start with 200 and add $250/100 \times 200 = 500$, for a total of $200 + 500 = 700$. A) 450 B) 500 C) 600 D) 700	24. D
25. If $a \bigstar b = 2a + b^4$, then $8 \bigstar 2 = 2 \times 8 + 2^4 = 16 + 16 = 32 = 2^5$. A) 2 B) 2^4 C) 2^5 D) 2^8	25. C

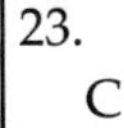

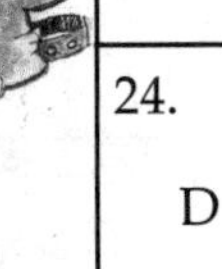

Go on to the next page ⟹ 8

26. The average of an even number of consecutive integers is always 0.5 more than an integer. Of the 4 choices, only 75 has this property (75 ÷ 6 = 12.5).

A) 67 B) 72 C) 75 D) 86

26. C

27. The numbers are _22, where _ is 1,4,5,6,7,8, or 9, and 2_2 or 22_, where _ is 0,1,4,5,6,7,8, or 9.

A) 23 B) 24 C) 26 D) 27

27. A

28. Let's find a set of numbers that will work. If there are 3 crocodiles, there are 9 cats, 45 koalas, and 20 cobras. The ratio of cobras to crocodiles is 20:3.

A) 4:3 B) 20:3 C) 45:4 D) 135:4

28. B

29. If 2 sides are 13, the 3rd side is any length from 1 to 20. If 13 is the 3rd side, the equal sides can be from 7 to 20. That's 20 + 14 − 1. (We counted all 13s twice.)

A) 14 B) 19 C) 32 D) 33

29. D

30. Among 120 athletes, there are 110 who play at least one of the 3 sports. Removing the 5 that play all 3 leaves 105, including baseball played by 52 others, basketball played by 47 others, and soccer played by 36 others. Those 3 numbers add to 135; exactly 30 must be counted twice.

A) 25 B) 30 C) 35 D) 40

30. B

31. If 9 of the friends write -7 and one of the friends writes 7, the product of their 10 integers is -7^{10}.

A) $(-7)^{10}$ B) $(-7)^9$ C) 0 D) -7^{10}

31. D

32. Each day Flo eats 1/50 of the food, so in 30 days she eats 30/50 of it. That means Jet eats 20/50 of the food in 30 days, or 1/75 of the food each day. Alone, Jet would last 75 days.

A) 10 B) 40 C) 70 D) 75

32. D

33. Powers have 1s digit cycle 8,4,2,6,8,4, ... The 1s digit is 6; 6 ÷ 5 = 1R1.

A) 1 B) 2 C) 3 D) 4

33. A

34. Since 36/100 = 9/25, the fewest coins I could have is 25. If I have 25, my dad has 1359. Since 1359 ÷ 25 = 54R9, the remainder is 9.

A) 1 B) 3 C) 9 D) 36

34. C

35. The product is odd if all 3 integers are odd, which has probability 1/2 × 9/19 × 8/18 = 2/19. The probability the product is even is 1−2/19.

A) 2/19 B) 1/8 C) 7/8 D) 17/19

35. D

The end of the contest  **8**

Visit our Website at http://www.mathleague.com

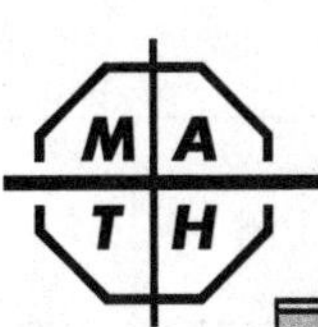

EIGHTH GRADE MATHEMATICS CONTEST

Math League Press, P.O. Box 17, Tenafly, New Jersey 07670-0017

Information & Solutions

Tuesday, February 16 or23, 2021

8

Contest Information

- **Solutions** Turn the page for detailed contest solutions (written in the question boxes) and letter answers (written in the *Answer Column* to the right of each question).

- **Scores** Please remember that *this is a contest, and not a test*—there is no "passing" or "failing" score. Few students score as high as 28 points (80% correct); students with half that, 14 points, *deserve commendation!*

- **Answers and Rating Scales** Turn to page 147 for the letter answers to each question and the rating scale for this contest.

Answers

1. $(111 \times 2) \times (111 \times 2) \times (111 \times 2) = 111 \times 111 \times 111 \times (2 \times 2 \times 2)$.

   A) 2 B) 2×3 C) 222 D) 2^3

   1. D

2. Since $123456789 \times 0.00001 = 1234.56789$, the hundredths digit is 6.

   A) 3 B) 4 C) 5 D) 6

   2. D

3. Since 1 ℓ = 1000 ml, 2 ℓ = 2000 ml.

   A) 20 B) 200 C) 2000 D) 20 000

   3. C

4. The greatest 4-digit integer with 3 different digits is 9987. The sum of its digits is 33.

   A) 30 B) 31 C) 32 D) 33

   4. D

5. Since $99 \div 3 = 33$, 33 are divisible by 3 and 66 are not.

   A) 33 B) 34 C) 66 D) 67

   5. C

6. Jan hiked 18 km in 3 hrs. Her average speed was 18 km/3hrs.

   A) 6 km/hr. B) 9 km/hr. C) 12 km/hr. D) 18 km/hr.

   6. A

7. The multiples of 8 that are factors of 64 are 8, 16, 32 and 64.

   A) 3 B) 4 C) 6 D) 8

   7. B

8. The greatest 3-digit perfect square is 961 and the greatest 2-digit perfect square is 81. Their difference is 880.

   A) 870 B) 880 C) 890 D) 900

   8. B

9. 10% of 100 = 10 = 20 − 10 = (10% of 200) − 10.

   A) 20 B) 110 C) 120 D) 200

   9. D

10. Consecutive multiples of 2 in order are 2, 4, 6, 8, 10, 12, 14, ... The 10th digit is a 4.

    A) 1 B) 2 C) 3 D) 4

    10. D

11. Today Al bought 36 hats. Yesterday he had 18 hats and today he has a total of 54 hats. Tomorrow Al buys 162 hats for a total of 216 hats.

    A) 108 B) 144 C) 162 D) 216

    11. D

12. The sum of 5 integers is divisible by 20. The average is divisible by $20 \div 5$.

    A) 4 B) 5 C) 20 D) 25

    12. A

13. Ed is late to work 5 days a month. If he works all 30 days in April, the probability Ed gets to work on time on a given day is 25/30.

    A) $\frac{1}{6}$ B) $\frac{1}{2}$ C) $\frac{2}{3}$ D) $\frac{5}{6}$

    13. D

Go on to the next page ⟹ 8

14. The number of slices in all 6 pizzas must be a multiple of 8. With 4 slices per pizza, there would be 24 slices and 24 is a multiple of 8.
    A) 2 B) 3 C) 4 D) 6
    14. C

15. Since 1.25 = 5/4, its reciprocal is 4/5.
    A) 0.125 B) 0.14 C) 0.4 D) 0.8
    15. D

16. If the difference between 2 prime numbers is a prime, the integers could be 2 and 5. Their sum is 2 + 5 = 7.
    A) 5 B) 7 C) 8 D) 9
    16. B

17. Three consecutive integers whose sum is 180 are 59, 60, and 61. The sum of the two smallest is 119.
    A) 118° B) 119° C) 120° D) 121°
    17. B

18. May painted 6 pots in 2 days or 3 pots in 1 day. Elsa and May together painted 10 pots in 3 days. May painted 9 of the pots, so Elsa painted 1 in 3 days. It would take Elsa 3×16 days to paint 16 pots.
    A) 12 B) 24 C) 36 D) 48
    18. D

19. Number of diagonals from a point is 3 less than the number of sides.
    A) octagon B) parallelogram C) hexagon D) pentagon
    19. C

20. The difference between the squares of 13 and 12 is 169 − 144 = 25. The sum of these squares is 169 + 144 = 313.
    A) 221 B) 265 C) 313 D) 365
    20. C

21. The product of 1, 2, 3, 4, 5, 6, 7, 8, and 9 is divisible by each prime less than 10. These primes are 2, 3, 5, and 7.
    A) 4 B) 5 C) 9 D) 13
    21. A

22. Diameter ÷ circumference = $d \div (\pi d) = 1/\pi$.
    A) $\frac{1}{\pi}$ B) $\frac{2}{\pi}$ C) $\frac{1}{2\pi}$ D) $\frac{\pi}{2}$
    22. A

23. The primes from 27 to 50 are 29, 31, 37, 41, 43, and 47. All other integers from 27 to 50 are divisible by an integer from 2 to 25.
    A) 17 B) 18 C) 19 D) 20
    23. B

24. If the length is 3 and the width 2, then the perimeter is 10. The ratio of the perimeter to the width is 10:2 = 5:1.
    A) 3:2 B) 5:2 C) 5:1 D) 9:4
    24. C

25. Since $6^3 = 2^3 \times 3^3$ and $8^3 = 2^3 \times 2^6$, the least multiple of 6^3 that is a multiple of 8^3 is $2^3 \times 3^3 \times 2^6$.
    A) 2×6^3 B) $2^3 \times 6^3$ C) $2 + 6^3$ D) $2^6 \times 6^3$
    25. D

Go on to the next page 8

26\. The value of 25 nickels and 15 dimes is 275¢. Each stack has the same value, so the number of stacks must be a factor of 275. The least possible is 5.

A) 3 B) 4 C) 5 D) 10

26. C

27\. The integers from 10 to 49 for which both digits are factors are 11, 12, 15, 22, 24, 33, 36, 44, and 48.

A) 9 B) 13 C) 17 D) 18

27. A

28\. $2^{2021} + 2^{2020} = (2^1 \times 2^{2020}) + (1 \times 2^{2020}) = 3 \times 2^{2020}$.

A) 2×2^{2020} B) 3×2^{2020} C) 4×2^{2020} D) $2^{2020} \times 2^{2020}$

28. B

29\. Writing ones digits only for powers: 9+1+9+1+9+1+9+1+9 = 49, so it's a 9.

A) 9 B) 7 C) 1 D) 49

29. A

30\. The probability that Aida chooses an A 1st is 1/2. The probability she chooses an I 2nd is 1/3. The probability she chooses a D 3rd is 1/2. Multiply these together to get 1/12.

A) $\frac{1}{16}$ B) $\frac{1}{12}$ C) $\frac{1}{8}$ D) $\frac{1}{4}$

30. B

31\. The last runner began running 199×15 seconds = 2985 seconds after the first runner began. Since both the first and the last runner ran for the same amount of time, the last runner stopped 2985 seconds after the first runner stopped running.

A) 2985 sec. B) 5970 sec. C) 8940 sec. D) 8955 sec.

31. A

32\. The first year I earned \$1000 interest. The second year I began with \$11 000 and earned \$1100 interest. The third year I began with \$12 100 and earned \$1210 interest for a total of \$1000+\$1100+\$1210 = \$3310.

A) \$3000 B) \$3100 C) \$3310 D) \$3641

32. C

33\. Multiply each of the 3 divisors of 25 by each divisor of 21^{100}.

A) 2×101^2 B) 3×101^2 C) 5×101^2 D) 25×101^2

33. B

34\. If the sum of the measures of the 2 smallest angles is 102°, the largest angle is 78°. If the sum of the measures of the 2 largest angles is 130°, the middle angle is 52°. The difference between them is 26°.

A) 2° B) 26° C) 28° D) 32°

34. B

35\. For each pair of consecutive 2-digit positive integers, the remainder when you square the larger integer and divide it by the smaller integer is 1. Since there are 89 pairs, the sum of the remainders for all such pairs is 89.

A) 44 B) 45 C) 89 D) 90

35. C

The end of the contest **8**

Visit our Website at http://www.mathleague.com

Algebra Course 1 Solutions

•••••••••••••••••••••••••

2016-2017 through 2020-2021

ALGEBRA COURSE 1 CONTEST

Math League Press, P.O. Box 17, Tenafly, New Jersey 07670-0017

Information & Solutions

Spring, 2017

Contest Information

- **Solutions** Turn the page for detailed contest solutions (written in the question boxes) and letter answers (written in the *Answer Column* to the right of each question).

- **Scores** Please remember that *this is a contest, and not a test*—there is no "passing" or "failing" score. Few students score as high as 24 points (80% correct); students with half that, 12 points, *deserve commendation!*

- **Answers and Rating Scales** Turn to page 148 for the letter answers to each question and the rating scale for this contest.

2016-2017 ALGEBRA COURSE 1 CONTEST SOLUTIONS — *Answers*

1. Whenever x is non-negative, $\sqrt{x}$ is a real number.

   A) $-1 < x < 1$ B) $x \geq 0$ C) $x < 0$ D) $x > 0$

   Answer: 1. B

2. $(j + k)^2 = (j + k)(j + k) = j^2 + \mathbf{2jk} + k^2$.

   A) 0 B) jk C) $2jk$ D) $j + k$

   Answer: 2. C

3. If t, u, b, and a are consecutive multiples of 2 and $t < u < b < a$, then the tuba band played $t + (t + 2) + (t + 4) + (t + 6) = (4t + 12)$ tunes.

   A) $t + 4$ B) $4t + 4$ C) $t + 12$ D) $4t + 12$

   Answer: 3. D

4. If $xy = 24$ when x is 3, then $3y = 24$. Thus, $y = 8$ and $y^2 = 64$.

   A) 72 B) 64 C) 16 D) 8

   Answer: 4. B

5. If $3 + x = 7$, then $7 = 3 + x$.

   A) $x + 3$ B) $x - 3$ C) $3x$ D) $3 - x$

   Answer: 5. A

6. The greatest common factor of $24x$ and $12x$ is $12x$.

   A) $2x$ B) $4x$ C) $6x$ D) $12x$

   Answer: 6. D

7. $\dfrac{1}{x^2 - 36}$ is undefined when $x^2 - 36 = 0$. This occurs if $x = 6$ or -6.

   A) 2 B) 3 C) 4 D) 6

   Answer: 7. A

8. Since $y^2 \geq 0$, if $x^3 = y^2$, then $x^3 \geq 0$. It follows that $x \geq 0$.

   A) non-negative B) even C) a perfect square D) odd

   Answer: 8. A

9. If $2x^2 - 9$ is divided by $x^2 - 9$, the quotient is 2 and the remainder is 9.

   A) -9 B) 0 C) 9 D) x^2

   Answer: 9. C

10. The sum of my slowest and fastest speeds is $(x^2 + 10x + 25)$ km/hr. If my fastest speed is $(x^2 + 25)$ km/hr., then my slowest speed is $(x^2 + 10x + 25) - (x^2 + 25) = 10x$ km/hr.

    A) x km/hr. B) $5x$ km/hr. C) $10x$ km/hr. D) $25x$ km/hr.

    Answer: 10. C

11. $x^1 \times x^2 \times x^3 \times x^4 = x^{1+2+3+4} = x^{10} = x^5 \times x^5$.

    A) x^{21} B) x^5 C) x^4 D) 1

    Answer: 11. B

12. $(s - e) - (e - s) = s - e - e + s = 2s - 2e$.

    A) 0 B) $s - 2e$ C) $2s - 2e$ D) $2s - e$

    Answer: 12. C

Go on to the next page ➡ **A**

2016-2017 ALGEBRA COURSE 1 CONTEST SOLUTIONS	*Answers*
13. Since $2^{2x} = (2^x)^2 = (2^2)^x = (4)^x = (2 \times 2)^x$, choice A isn't equivalent. A) 2×2^x B) $(2 \times 2)^x$ C) $(2^2)^x$ D) $(2^x)^2$	13. A
14. $\lvert x^2 - 2x + 1 \rvert = \lvert (x-1)^2 \rvert = \lvert (1-x)^2 \rvert$. A) $\lvert 1 - x \rvert$ B) $\lvert x - 1 \rvert$ C) $\lvert (1-x)^2 \rvert$ D) $x^2 + 2x + 1$	14. C
15. If x and y are integers with $y < 2017$, $x > 0$, and $x^2 < y$, $x < \sqrt{y} \le 44$; $x = 1, 2, 3, \ldots, 44$. A) 44 B) 45 C) 88 D) 89	15. A
16. When I play darts, I score 5 points for each of my 1st 10 bullseyes and 8 points for each bullseye made after that. I get 50 points for the 1st 10 and $8b$ for the other bullseyes. That's a total of $50 + 8b$ points. A) $8b$ B) $50 + 3b$ C) $50 + 8b$ D) $50 + 15b$	16. C
17. If the numerical area of a square is s^2, then its numerical perimeter is $4s$. If $4s = s^2 + 3$, $s^2 - 4s + 3 = 0$. Possible values of s are 1 and 3. A) 5 B) 9 C) 12 D) 21	17. D
18. Parallel lines have = slopes; the product (if defined) is a perfect square. A) 181 B) 183 C) 187 D) 289	18. D
19. The l.c.m. of $(x+1)(x-1)$ and $(x-1)^2$ is $(x+1)(x-1)^2$. A) $(x^2-1)(x+1)$ B) $(x^2-1)(x-1)^2$ C) $(x+1)(x-1)^2$ D) $(x+1)(x-1)^3$	19. C
20. If $f(5) = 5$, then $f(f(5)) = f(5) = 5$. A) 1 B) 5 C) 10 D) 25	20. B
21. Anna rode exactly 1 km. For each tire revolution she travels πd km; 80 revolutions is $80\pi d$ km. Hence, $80\pi d = 1$ km. Solve for d to get choice A. A) $\frac{1}{80\pi}$ km B) $\frac{80}{\pi}$ km C) $\frac{\pi}{80}$ km D) 80π km	21. A
22. $2^x + 2^x + 2^x + 2^x = 4 \times 2^x = 2^2 \times 2^x = 2^{2+x}$. A) 2^{4x} B) 2^x C) 2^{2+x} D) 2^{2x}	22. C

Go on to the next page ⟹ **A**

23. It's true for x = 0, 5, 6, 7, 8, 9, and 10; for $x = 1, 2, 3$, and 4, $\frac{(x-1)(x-2)(x-3)(x-4)}{(x-1)(x-2)(x-3)(x-4)} = 1$ is undefined.

    A) 7 B) 8 C) 9 D) 10

    23. A

24. Since 2 hours = 7200 seconds, and each second consists of 15 frames, I need $15 \times 7200 = 108\,000$ frames to make a 2-hour stop-motion movie.

    A) 30 B) 1800 C) 72 000 D) 108 000

    24. D

25. Since $\frac{n!}{(n-2)!} = n(n-1)$, we have $n^2 - n - 8010 = 0$. The only positive solution of this is $n = 90$.

    A) 60 B) 70 C) 80 D) 90

    25. D

26. Let v = number of veggies originally. Then $4v$ = number of noodles originally. After eating 50 noodles and 8 veggies, the new ratio is 5:2 or $(4v - 50):(v - 8)$. Setting these ratios equal and simplifying, $(8v - 100) = (5v - 40)$. Thus, $v = 20$ and $(4v - 50) + (v - 8) = 42$.

    A) 42 B) 58 C) 98 D) 100

    26. A

27. The real solutions of $\frac{(x^2-25)(x^2-16)(x^2-9)(x^2-4)(x^2-1)}{(x+1)(x+2)(x+3)(x+4)(x+5)} = 0$ are 5, 4, 3, 2, and 1.

    A) 0 B) 5! C) -(5!) D) $5! \times 5!$

    27. B

28. Darth fills egg cartons at a constant rate of 24 cartons every quarter of an hour. Luke fills 72 egg cartons in that same time. In 8 hours, Darth and Luke can fill $8 \times 4 \times (72 + 24) = 3072$ cartons.

    A) 576 B) 768 C) 2304 D) 3072

    28. D

29. I need to make up a secret 7-digit passcode. All digits must be different; no consecutive numbers can be next to each other; no even digit can be next to an even digit; and no odd digit can be next to an odd digit. The greatest possible such passcode is 9638527.

    A) 6 B) 7 C) 8 D) 9

    29. C

30. If $n = 4$, then $((4^2)^2)^2 = 4^8 = 2^{16}$. The 9 even powers of 2 from 0 to 16 are perfect squares.

    A) 16 B) 9 C) 8 D) 4

    30. B

The end of the contest 

Visit our Web site at http://www.mathleague.com

Math League Press, P.O. Box 17, Tenafly, New Jersey 07670-0017

Information & Solutions

Spring, 2013

Contest Information

- **Solutions** Turn the page for detailed contest solutions (written in the question boxes) and letter answers (written in the *Answer Column* to the right of each question).

- **Scores** Please remember that *this is a contest, and not a test*—there is no "passing" or "failing" score. Few students score as high as 24 points (80% correct); students with half that, 12 points, *deserve commendation!*

- **Answers and Rating Scales** Turn to page 149 for the letter answers to each question and the rating scale for this contest.

1. If $a+l+g+e+b+r+a = 38$ and $a+l+g+e+b+a = 21$, subtract the 2nd equation from the 1st to get $r = 17$.
   A) 1 B) 10 C) 17 D) 59
   Answer: 1. C

2. Since 1 km = 1000 m, x km = $1000x$ m. Therefore, Harald traveled $1000x$ m.
   A) $100x$ B) $1000x$ C) $10\,000x$ D) $1\,000\,000x$
   Answer: 2. B

3. $a+2+2a+4-3a+6-4a-8 = -4a+4$.
   A) $4a-8$ B) $4a+4$ C) $-4a-8$ D) $-4a+4$
   Answer: 3. D

4. $b^4-81 = (b^2+9)(b^2-9) = (b^2+9)(b+3)(b-3)$.
   A) $b-3$ B) $b+3$ C) $b-9$ D) b^2-9
   Answer: 4. C

5. Multiplying both sides of the given inequality by s^{12}, we get $s > 1$.
   A) 2 B) -2 C) $\frac{1}{2}$ D) $-\frac{1}{2}$
   Answer: 5. A

6. If $60s + 90t = 120s$, then $30t = 60s - 60t$. Thus, $0.5t = s - t$.
   A) $0.5t$ B) $2t$ C) $15t$ D) $30t$
   Answer: 6. A

7. $\frac{x^2}{y^2} \div \frac{x^1}{y^1} = \frac{x^2}{y^2} \times \frac{y^1}{x^1} = \frac{x}{y}$.
   A) xy B) $\frac{x}{y}$ C) $\frac{y}{x}$ D) $\frac{x^3}{y^3}$
   Answer: 7. B

8. $(n-4)^2 - (4-n)^2 = (n-4)^2 - (n-4)^2 = 0$.
   A) 0 B) $-16n$ C) $16n$ D) $2n^2 - 16n$
   Answer: 8. A

9. If p is a 3-digit prime number, then choices A and D are divisible by 2 and choice B is divisible by 3. If $p = 103$, then $p - 2 = 101$ is prime.
   A) $2p$ B) $3p$ C) $p-2$ D) $p+3$
   Answer: 9. C

10. My dog will catch up to the slowest member of the band in s seconds. If $s^2 - 16s + 64 = 0$, then $(s-8)^2 = 0$. My dog will catch up in 8 seconds.
    A) 2 B) 4 C) 6 D) 8
    Answer: 10. D

11. $(z^2 \times z^4 \times z^6 \times z^8)^2 = (z^{20})^2 = z^2 \times z^{38}$.
    A) z^{766} B) z^{40} C) z^{38} D) z^{20}
    Answer: 11. C

12. The line $3x - 4y = 5$ is parallel to the line $3x - 4y = -5$.
    A) $2x - 7y = 5$ B) $4y - 3x = 5$ C) $3x - 6y = 7$ D) $2x - 4y = 6$
    Answer: 12. B

Go on to the next page ➠ **A**

13. There are $4x$ boys and $5y$ girls in my class, and the ratio of boys to girls is 4:5. Therefore, $4x:5y = 4:5$ and $x:y = 1:1$.
A) 4:5 B) 5:4 C) 1:1 D) 16:25
Answer 13. C

14. If the average of r, s, and t is 20, their sum is 60. If the average of r and t is 24, their sum is 48. Thus, the value of s is $60 - 48 = 12$.
A) 4 B) 12 C) -4 D) -12
Answer 14. B

15. Mrs. Robinson's son begs her for a cookie every $42m$ minutes, and her dog begs her for a cookie every $12m$ minutes, where m is a prime greater than 7. The next time they both beg for a cookie at the same time will be the l.c.m. of $42m$ and $12m$ minutes later. The l.c.m of $42m$ and $12m$ is $84m$.
A) 30 B) $54m$ C) $84m$ D) $504m^2$
Answer 15. C

16. If $a \Diamond b = 2a^2 + 3ab$, then $4 \Diamond 5 = 2(4^2) + 3(4)(5) = 92$.
A) 18 B) 24 C) 60 D) 92
Answer 16. D

17. ($18x$ km/hr.)(1000 m/km)(1 hr./60 min.)(1min./60 sec.) = $5x$ m/sec.
A) $5x$ B) $9x$ C) $18x$ D) $36x$
Answer 17. A

18. If n is even, choice A is an odd number. If n is odd, choice B is an odd number. If $n = 3$, choice C has no factor of 3.
A) $(n+1)(n+3)(n+5)$ B) $(n+2)(n+4)(n+6)$
C) $(n+2)(n+5)(n+8)$ D) $(n+1)(n+3)(n+8)$
Answer 18. D

19. $|10-4x| = 5$ if and only if $10 - 4x = 5$ or $10 - 4x = -5$; $x = 5/4$ or $15/4$.
A) 1.25 B) 3.75 C) 5 D) 10
Answer 19. C

20. If $2^x \times 2^{4x} \times 2^{9x} = 2^{14x} = 2^y$, then $y = 14x$.
A) $2x^3$ B) $6x$ C) $6x^3$ D) $14x$
Answer 20. D

21. Alan has 5 complete asleep and then awake periods, plus one more asleep period. That's $5(12m) + 8m = 68m$ minutes.
From 11 P.M. until 4:06 A.M. is 306 minutes. Since $68m = 306$, $m = 4.5$.
A) 4.25 B) 4.5 C) 5.125 D) 6.375
Answer 21. B

22. Since powers of 8 can only end in 2, 4, 6, or 8, the remainder when 2018^x is divided by 10 could not be 0.
A) 4 B) 6 C) 8 D) 0
Answer 22. D

Go on to the next page ⟹ A

2017-2018 ALGEBRA COURSE 1 CONTEST SOLUTIONS	Answers
23. Since $\frac{1}{a}+\frac{1}{b} = \frac{a+b}{ab}$ and $a+b=8$, $\frac{8}{ab} = 4$ and $ab = 2$. A) 2 B) 6 C) 12 D) 32	23. A
24. Let the distance around the world be $3600d$ km. It takes $4d$ hours for the 1st trip, $3d$ hours for the 2nd trip, and $2d$ hours for the 3rd. The average rate for all 3 laps is $10\,800d$ km/$9d$ hr. = 1200 km/hr. A) 1200 B) 1300 C) 1400 D) 1500	24. A
25. Multiply the average, $(1+2n)/2$, by the number of integers, $2n$. A) $2n^2+n$ B) $2n^2+1$ C) $2n^2+n+1$ D) $2n^2+2n+2$	25. A
26. Subtract $2(3p-4q=5)$ from $4(2p+7q=11)$ to get $2p+36q=34$. A) 6 B) 16 C) 34 D) 55	26. C
27. If the average age of the 5 youngest siblings is 6, the sum of their ages is 30. If the average age of the 5 oldest siblings is 18, the sum of their ages is 90. The difference between these sums, 90 – 30, is the difference in the ages of the 4 oldest and the ages of the 4 youngest. Dividing this difference by 4, the difference in their average ages is 15. A) 8 B) 10 C) 15 D) 19	27. C
28. Squaring, $a^2-6a+9 = b^2+12b+36$; $b^2+12b+24 = a^2-6a-3$. A) a^2-6a-3 B) a^2-6a+9 C) $a^2+6a-12$ D) a^2+6a	28. A
29. If g = the number of goats hired 2 years ago, then $g+6$ = the number of goats hired last year and $g-26$ = the number of sheep hired last year. The total number of animals hired last year was $2g-20$. We have $(g+6)/(2g-20) = 3/4$. Solving, $g = 42$. Thus, $n = 2(42)-20 = 64$. A) 18 B) 22 C) 48 D) 64	29. D
30. To count divisors, multiply exponents after adding 1 to each exponent in prime factorization. If $x = 720 = 2^4 \times 3^2 \times 5^1$, x has $5{\times}3{\times}2 = 30$ divisors. A) 9 B) 16 C) 25 D) 31	30. A

The end of the contest 

Visit our Website at http://www.mathleague.com

ALGEBRA COURSE 1 CONTEST

Math League Press, P.O. Box 17, Tenafly, New Jersey 07670-0017

Information & Solutions

Spring, 2019

Contest Information

- **Solutions** Turn the page for detailed contest solutions (written in the question boxes) and letter answers (written in the *Answer Column* to the right of each question).

- **Scores** Please remember that *this is a contest, and not a test*—there is no "passing" or "failing" score. Few students score as high as 24 points (80% correct); students with half that, 12 points, *deserve commendation!*

- **Answers and Rating Scales** Turn to page 150 for the letter answers to each question and the rating scale for this contest.

Solutions	Answers
1. If $a = 2$, $r = 0$, $t = 1$, and $s = 9$, then $s + t + a + r + t = 9 + 1 + 2 + 0 + 1 = 13$. A) 0 B) 12 C) 13 D) 21	1. C
2. There were a ants in my ant farm. They have $6a$ legs. After 3 ants leave, the remaining ants have $6a - 18 = 6(a - 3)$ legs. A) $6a - 3$ B) $6(a - 3)$ C) $6a - 3a$ D) $a^6 - 3$	2. B
3. Regroup: $(6x^2 + 2x^2 + 4x^2) + (4x + 2x + 6x) - (5+3+1+3+5)$. A) $36x - 17$ B) $24x - 9$ C) $12x^2 + 12x - 12$ D) $12x^2 + 12x - 17$	3. D
4. $(x - y)(x + y) = x^2 + xy - xy - y^2 = x^2 - y^2$. A) $x^2 - y^2$ B) $x^2 - 2xy + y^2$ C) $x^2 + 2xy + y^2$ D) $x^2 + y^2$	4. A
5. $(x - y)(x + y)(x - y) = (x^2 - y^2)(x - y) = x^3 - x^2y - xy^2 + y^3$. A) $x^3 - y^3$ B) $x^3 - x^2y - xy^2 + y^3$ C) $x^3 + y^3$ D) $x^3 + x^2y + xy^2 + y^3$	5. B
6. Since $-s^2 \leq 0$ for all real values of s, $-s^2 - 1 < 0$ for all real values of s. A) $-s^3 - 1$ B) $(-s)^3 - 1$ C) $-s^2 - 1$ D) $(-s)^2 - 1$	6. C
7. The integer solutions of $(x^2 - 1)(x^2 - 2)(x^2 - 3)(x^2 - 4) = 0$ are ± 1, ± 2. A) 2 B) 4 C) 6 D) 8	7. B
8. If x, y, and z are distinct prime numbers, the least common multiple of $x^2y^3z^4$ and $x^4y^3z^2$ must contain the highest power of each prime. A) $x^8y^9z^8$ B) $x^6y^6z^6$ C) $x^4y^3z^4$ D) $x^2y^3z^2$	8. C
9. $((x^3 + x^3) \times x^3)^3 = (2x^3 \times x^3)^3 = (2x^6)^3 = 2^3x^{18} = 8x^{18}$. A) $2x^{18}$ B) $8x^{18}$ C) $8x^{27}$ D) x^{54}	9. B
10. In my jar, there are $3b$ red beans, 5b green beans, $6b$ orange beans, for a total of $14b$ beans. If $b = 3$, the total number of beans would be 42. A) 35 B) 42 C) 60 D) 90	10. B
11. $2x - 2.5 = \pm 4$, so $x = 3.25$ or -0.75. The sum of the solutions is 2.5. A) 2 B) 2.5 C) 3.75 D) 4	11. B
12. The roots of $(x - 7)(x + 4) = 0$ are 7 and -4. Their difference is 11. A) 3 B) 4 C) 7 D) 11	12. D

Go on to the next page ⟫➜ **A**

13. Today Li turned 42 and Mae turned 8. In x years, we want $42 + x = 3(8 + x)$. Solving, $x = 9$. Therefore, Mae will be 17.

A) 9 B) 17 C) 26 D) 51

13. B

14. Three crates contain $3b$ boxes, and three boxes contain $3bp$ packages. If each package holds 4 bulbs, three crates contain $12bp$ bulbs.

A) $12bp$ B) $\frac{3bp}{4}$ C) $\frac{4bp}{3}$ D) $\frac{bp}{12}$

14. A

15. Before the wave hit, $a = 3b$. After the wave hit, $(a - 3)/(b + 1) = 5/2$. Combining these equations, $(3b - 3)/(b + 1) = 5/2$. Simplifying, $6b - 6 = 5b + 5$. Solving, $b = 11$. Since $a = 3b$, $a = 33$. Thus, Avi had built 33 sand castles before the wave hit.

A) 11 B) 12 C) 30 D) 33

15. D

16. If $135 \times (46 + 2) = (135 \times 46) + 270 = a + 270$.

A) $a + 2$ B) $a + 92$ C) $a + 94$ D) $a + 270$

16. D

17. If $3x + 8y = 21$ and $8x + 3y = 23$, $11x + 11y = 44$ and $x + y = 4$.

A) 2 B) 4 C) 11 D) 22

17. B

18. If the hands on a circular clock start at midnight, 1000 hours later is 83 full times around and then one-third more, which is 4 hours.

A) 2 B) 4 C) 8 D) 12

18. B

19. If $x = 3$, the value of $|20 - 7x|$ is 1.

A) 1 B) 2 C) 3 D) 6

19. A

20. If Sy can shovel snow from half of a driveway in 2 hours, and Ty can shovel snow from one quarter of the driveway in 2 hours, together they shovel three-quarters of the driveway in 120 minutes or one quarter in 40 minutes or four-quarters in 160 minutes.

A) 120 B) 160 C) 180 D) 360

20. B

21. Of the bottles that Viola collects, 80% are green. Of the green bottles, 30% held perfume 45% held spices. Thus, 25% of the green bottles held pills. Since 25% of 80% is 20%, and 20% of her bottles is 25, 100% of her bottles is 125.

A) 75 B) 100 C) 120 D) 125

21. D

22. Clearing fractions, $2x^2 - y + 3x^2 = 4$; $y = 5x^2 - 4$.

A) $4 - x^2$ B) $4 + x^2$ C) $5x^2 - 4$ D) $4 - 5x^2$

22. C

Go on to the next page ➡ **A**

23. Don ate $x - 27$ cherries and Juan ate $x - 11$ cherries. Since $x - 27 \geq 10$ and $x - 11 \geq 10$, $x \geq 37$. In addition, $x - 11 + x - 27 \leq x - 1$, so $x \leq 37$. Therefore, $x = 37$.

A) 37 B) 38 C) 39 D) 49

23. A

---

24. If we subtract a and b from 200, we subtract the pets with scales **and** gills twice. Adding them back once, we have $200 - a - b + c$ with neither.

A) $200 - a - b$ B) $200 - c$ C) $200 - a - b - c$ D) $200 - a - b + c$

24. D

---

25. Since $xy = 144$ and $y = 3x - 6$, $3x^2 - 6x = 144$. Hence, $x^2 - 2x - 48 = 0$. Thus, $(x + 6)(x - 8) = 0$ and $x = -6$ or 8. Since $y < x$, $x = -6$.

A) 18 B) 8 C) -6 D) -24

25. C

---

26. If $x + y = 2$, $y = 2 - x$ and $20x + 50y = 20x + 50(2 - x) = 100 - 30x < 100$. Similarly, $x = 2 - y$ and $20x + 50y = 40 + 30y > 40$.

A) 35 B) 65 C) 105 D) 140

26. B

---

27. If the border's width is y, the area of the border is $2y(2x) + 2y(3x) + 4y^2 = 4y^2 + 10xy$. We are given the area is $14x^2$, so $4y^2 + 10xy = 14x^2$. Thus, $(2y + 7x)(y - x) = 0$. Since $y > 0$, $y = x$.

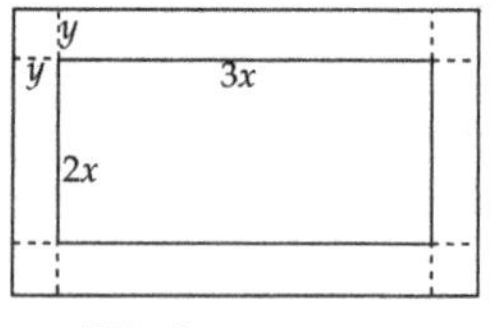

A) $0.5x$ m B) x m C) $1.5x$ m D) $2x$ m

27. B

---

28. 10^{2019} has 2020 digits. Subtracting 2019 from 10^{2019}, the result is 999999999 . . . 999997981. That's 2015(9)+7+9+8+1 =18160.

A) 2019 B) 18160 C) 18161 D) 18169

28. B

---

29. The percent of sugar is $(10x+20y)/(x+y)$. Set this equal to z and solve: $10x+20y = xz+yz$, so $(10 - z)x = (z - 20)y$ and $x/y = (z - 20):(10 - z)$.

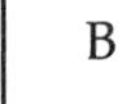
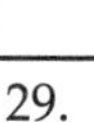

A) $(20 - z):(z - 10)$ B) $(10 - z):(z + 20)$
C) $(z + 10):(20 - z)$ D) $(z + 20):(10 - z)$

29. A

---

30. If x, y, and z are prime, the whole-number divisors of the product xyz are 1, x, y, z, xy, xz, yz, and xyz. The product of these is $x^4y^4z^4$.

A) xyz B) $x^2y^2z^2$ C) $x^3y^3z^3$ D) $x^4y^4z^4$

30. D

The end of the contest **A**

Visit our Web site at http://www.mathleague.com

Math League Press, P.O. Box 17, Tenafly, New Jersey 07670-0017

Information & Solutions

A

Spring, 2020

Contest Information

- **Solutions** Turn the page for detailed contest solutions (written in the question boxes) and letter answers (written in the *Answer Column* to the right of each question).

- **Scores** Please remember that *this is a contest, and not a test*—there is no "passing" or "failing" score. Few students score as high as 24 points (80% correct); students with half that, 12 points, *deserve commendation!*

- **Answers and Rating Scales** Turn to page 151 for the letter answers to each question and the rating scale for this contest.

1. If $T = 1$, $O = 2$, and $1+2+S+S = 7$, then $2S = 4$, and $S = 2$.

   A) 2 B) 3 C) 3.5 D) 4

   Answer 1. A

2. If x is an integer, then the least possible value of $4x^2$ is obtained when $x = 0$.

   A) -4 B) 0 C) 4 D) 16

   Answer 2. B

3. $(c^{20})(c^2)(c^0) = c^{20+2+0} = c^{22}$.

   A) 0 B) c^0 C) c^{22} D) c^{40}

   Answer 3. C

4. I had g invited guests. Each invited guest brought 2 uninvited friends, for $2g$ additional people. Each person brought two gifts. Multiply total people $(g + 2g)$ by the number of gifts each brought, 2.

   A) $(g+2) \times 2$ B) $(g \times 2) + 2$ C) $(g + 2g) \times 2$ D) $(g + 2g) \times 2g$

   Answer 4. C

5. $4y(x-y) - (3x+2y)(x-y) = [4y - (3x+2y)](x-y) = [4y-3x-2y](x-y)$.

   A) $(6y + 3x)(x-y)$ B) $(6y - 3x)(x-y)$
   C) $(2y + 3x)(x-y)$ D) $(2y - 3x)(x-y)$

   Answer 5. D

6. $4x^2 + 3x + 2x^3 - 2x^2 - 3x - 4x^3 = (3x - 3x) + (4x^2 - 2x^2) + (2x^3 - 4x^3)$.

   A) 0 B) $2x^2 - 2x^3$ C) $2x^2 + 6x - 2x^3$ D) $2x^2 + 6x + 6x^3$

   Answer 6. B

7. If $\frac{3}{5}(2y) = \frac{4}{7}x$, multiply by 5/3: $2y = \frac{20}{21}x$. Divide by 2 to get $\frac{10}{21}x$.

   A) $\frac{10}{21}x$ B) $\frac{20}{21}x$ C) $\frac{21}{20}x$ D) $\frac{21}{10}x$

   Answer 7. A

8. $(x+2)(x-2)(x^2-4) = 0 = (x+2)^2(x-2)^2$, so x can be -2 or 2.

   A) 1 B) 2 C) 3 D) 4

   Answer 8. B

9. If $x > 5$ and prime, the l.c.m. of $2^2 5^1 x^2$ and $2^1 3^1 5^1 x^3$ is $2^2 3^1 5^1 x^3$.

   A) $10x$ B) $60x^3$ C) $60x^5$ D) $600x^5$

   Answer 9. B

10. Water time:walking time is 30 sec.:120 sec. = 1:4, so they spend 1/5 of the h hrs. in the water. Since h hrs. = $60h$ minutes, they will spend $60h/5$ min. = $12h$ min. in the water.

    A) $12h$ B) $24h$ C) $36h$ D) $48h$

    Answer 10. A

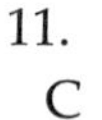

11. If $x = y+1$, then $(y+1)^2 - y^2 = 2y+1 = 39$; $y = 19$, $x = 20$.

    A) 39 B) 78 C) 380 D) 1521

    Answer 11. C

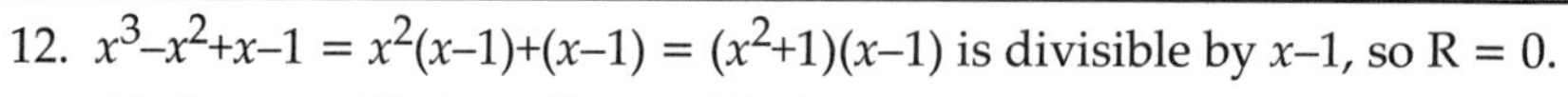

12. $x^3 - x^2 + x - 1 = x^2(x-1) + (x-1) = (x^2+1)(x-1)$ is divisible by $x-1$, so R = 0.

    A) 0 B) 1 C) x D) $2x$

    Answer 12. A

Go on to the next page ⟹ **A**

13. A $\perp$ line has negative reciprocal slope, so $y = -3x + b$. Substitute 0 for y in original line to find x-intercept -12; use to find new b.

    A) $y = -3x + 4$ B) $y = 3x - 36$ C) $y = \frac{1}{3}x + 4$ D) $y = -3x - 36$

    13. D

14. The sum of all solutions is $-b/a$, so for $4x^2 - 4x - 35 = 0$: $-(-4)/4 = 1$.

    A) -1 B) 0 C) 1 D) 4

    14. C

15. If $b^2 + 31 = g^4 = (g^2)^2 = (b + 1)^2$, then $b^2 + 31 = b^2 + 2b + 1$, so $b = 15$. Substitute $b = 15$ in $b + 1 = g^2$; $16 = g^2$ and $g = 4$.

    A) 4 B) 11 C) 12 D) 30

    15. A

16. Rate r m/min.$=60r$ m/hr.$=60r/1000$ km/hr. Divide distance k by rate to find time.

    A) $\frac{1000}{rk}$ B) $\frac{60k}{1000r}$ C) $\frac{1000k}{60r}$ D) $\frac{1000}{60rk}$

    16. C

17. $4^{2x} + 4^{2x} + 4^{2x} + 4^{2x} = 4(4^{2x}) = 4^{2x+1} = (2^2)^{2x+1} = 2^{4x+2}$.

    A) 2^{4x} B) 2^{4x+2} C) 4^{8x} D) 16^{2x}

    17. B

18. If $f(x) = 8x^2 - 2$, then $f(4) = 8(4)^2 - 2 = 126$. $f(-4)$ will also yield 126.

    A) $f(126)$ B) $f(8)$ C) $f(-2)$ D) $f(-4)$

    18. D

19. If $3x - 7 < 5$, $x < 4$. If $3x - 7 > -5$, $x > 2/3$. So $2/3 < x < 4$. 1, 2, and 3 fit.

    A) 1 B) 2 C) 3 D) 6

    19. C

20. Multiply $x^2+x+1 = 18$ by x to get $x^3+x^2+x = 18x$. The average is $18x/3$.

    A) $6x$ B) $9x$ C) $18x$ D) $36x$

    20. A

21. The Cones is an elite *a cappella* vocal group. Together, 4 Cones working at the same rate can set up every chair in the theater in 56 min. So it takes $4 \times 56 = 224$ Cone-min. of work to set up every chair. Divide 224 Cone-min. by 7 Cones to find that it will take 32 min. for 7 Cones working at the same rate to set up every chair in the theater.

    A) 24 B) 32 C) 48 D) 98

    21. B

22. $10^a = (0.01/100) \times 10^b = (10^{-2}/10^2) \times 10^b = 10^{-4} \times 10^b = 10^{-4+b}$, so $a = b - 4$.

    A) $b - 4$ B) $b - 2$ C) $b + 2$ D) $b + 4$

    22. A

Go on to the next page ⟫➡ **A**

23. The fish dragged his boat x km east, then $6x + 3$ km north, then $x + 8$ km east, then $2x - 2$ km south, then $2x + 8$ km west, then x^2 km south. The distance to the north must equal the distance to the south, so $6x+3 = 2x-2 + x^2$. Thus, $x^2 - 4x - 5 = 0 = (x - 5)(x + 1)$. The only possible value of x is 5, so substitute it in for all distances: $5 + 6(5)+3 + 5+8 + 2(5)-2 + 2(5)+8 + 5^2 = 102$.

A) 5 B) 30 C) 81 D) 102

23. D

24. I mixed 400 ml of 30% sugar lemonade and 200 ml of 40% sugar lemonade. That gave me 120 ml + 80 ml = 200 ml of sugar out of 600 ml of lemonade in the two batches. During the week 100 ml of pure water evaporated, leaving 200 ml sugar in 500 ml lemonade.

A) 30% B) 33% C) 35% D) 40%

24. D

25. If $\sqrt{xy} \times \sqrt{15} = \sqrt{3x^2} \times \sqrt{y}$, then $15xy = 3x^2y$, and $x = 5$.

A) 5 B) y C) $5y$ D) $5 + y$

25. A

26. $\dfrac{x^2 - 3x - 18 - x + 6}{(x-6)(x+3)} = \dfrac{x^2 - 4x - 12}{(x-6)(x-3)} = \dfrac{(x-6)(x+2)}{(x-6)(x-3)} = \dfrac{x+2}{x+3}$.

A) 1 B) $\dfrac{x+2}{x+3}$ C) $\dfrac{x-6}{x+3}$ D) $\dfrac{-2}{-(x-6)}$

26. B

27. $3(3x - 4y + 5z = 13) - 2(4x - 5y + 6z = 18) \Longrightarrow x - 2y + 3z = 3$.

A) 1 B) 3 C) 15 D) 28

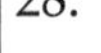

27. B

28. Between 1000 and 5000 knights were at the start. Each day 2/3 of the remaining knights fell or fled. Yesterday Saul lost his final 2 fellow knights. Work backward: Yesterday there were 3, the day before 9, before that 27, then 81, then 243, then 729, then 2187. 7 days.

A) 70 B) 14 C) 7 D) 6

28. C

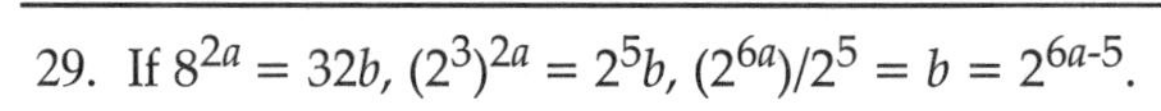

29. If $8^{2a} = 32b$, $(2^3)^{2a} = 2^5b$, $(2^{6a})/2^5 = b = 2^{6a-5}$.

A) 2^a B) $2^{6a/5}$ C) 2^{2a-3} D) 2^{6a-5}

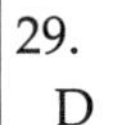

29. D

30. Find factors of 2: 1000/2=500, 1000/4=250, 1000/8=125, 1000/16=62.... There are 500+250+125+62+31+15+7+3+1=994 2's. $2^{994}=(2^2)^{497}=4^{497}$.

A) 250 B) 312 C) 330 D) 497

30. D

The end of the contest ✍ **A**

Visit our Website at http://www.mathleague.com

ALGEBRA COURSE 1 CONTEST

Math League Press, P.O. Box 17, Tenafly, New Jersey 07670-0017

Information & Solutions

A

Spring, 2021

Contest Information

- **Solutions** Turn the page for detailed contest solutions (written in the question boxes) and letter answers (written in the *Answer Column* to the right of each question).

- **Scores** Please remember that *this is a contest, and not a test*—there is no "passing" or "failing" score. Few students score as high as 24 points (80% correct); students with half that, 12 points, *deserve commendation!*

- **Answers and Rating Scales** Turn to page 152 for the letter answers to each question and the rating scale for this contest.

2020-2021 ALGEBRA COURSE 1 SOLUTIONS

Solutions	*Answers*
1. If $x = -2021$, only choices A and B are positive, and **A** is greatest. A) $-2021x$ B) $-2020x$ C) $2020x$ D) $2021x$	1. A
2. The letters in *MATHISFUN* represent the 9 digits 1, 2, . . . , 9 in that order. So *T* is 3, *I* is 5, *N* is 9, and *A* is 2. The code *TINA* is **3592**. A) 3492 B) 3582 C) 3592 D) 3594	2. C
3. If $2x + 5 = 13$, multiply by 4 to get $8x + 20 =$ **52**. A) 26 B) 29 C) 39 D) 52	3. D
4. If $8a = 12$ and $ab \neq 0$, divide by $8b$ to get the value of $\frac{a}{b} = \frac{12}{8} = \mathbf{\frac{3}{2}}$. A) $\frac{1}{2}$ B) $\frac{2}{3}$ C) $\frac{3}{2}$ D) 2	4. C
5. $2a - 3b + 4 - 9a + 13b - 15 = 2a - 9a - 3b + 13b + 4 - 15 = \mathbf{-7a + 10b - 11}$. A) $-7a - 10b - 11$ B) $-7a + 10b - 11$ C) $7a + 10b - 11$ D) $-7a + 10b + 11$	5. B
6. $(a + 2)(3a - 4) = 3a^2 - 4a + 6a - 8 = \mathbf{3a^2 + 2a - 8}$. A) $a^2 - 2a - 8$ B) $3a^2 - 2a - 8$ C) $3a^2 - 2a + 8$ D) $3a^2 + 2a - 8$	6. D
7. $x^2 - (x+4)(x-4) = x^2 - (x^2 - 4x + 4x - 16) = x^2 - (x^2 - 16) = x^2 - x^2 + 16 = \mathbf{16}$. A) -16 B) -8 C) 8 D) 16	7. D
8. If x is an integer, $125x^6 = 5^3x^6 = (5x^2)^3$ must be the **3rd** power of an integer. A) 2nd B) 3rd C) 5th D) 6th	8. B
9. The sum of the ages of my 3 children is 45 years. Consecutive odd integers $x + (x+2) + (x+4) = 45$, so $3x + 6 = 45$ and $x = 13$. Their ages are 13, 15, and **17**. A) 17 B) 19 C) 21 D) 23	9. A
10. L.c.d. ab: $\frac{1}{a} - \frac{1}{b} = \frac{b-a}{ab} = \frac{-(a-b)}{ab} = \frac{-10}{50} = \mathbf{-\frac{1}{5}}$. A) -5 B) $-\frac{1}{5}$ C) $\frac{1}{5}$ D) 5	10. B
11. If a and b are positive numbers, $\sqrt{2.56a^6b^{10}} = 1.6(a^6b^{10})^{1/2} = \mathbf{1.6a^3b^5}$. A) $1.28a^3b^5$ B) $1.28a^6b^{10}$ C) $1.6a^3b^5$ D) $1.6a^6b^{10}$	11. C
12. If today is Tuesday, $14n$ days from now (where n is a positive integer) will also be a Tuesday. Add 3 more days and it will be **Friday**. A) Wednesday B) Thursday C) Friday D) Saturday	12. C

Go on to the next page ➠ **A**

13. If 4% are vegan and 120 are not, 120 is 96% of the students.

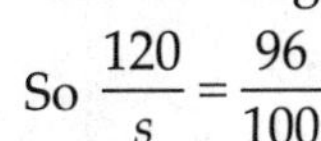

So $\frac{120}{s} = \frac{96}{100}$, $12000 = 96s$, and $s =$ **125**.

A) 123 B) 124 C) 125 D) 126

13. C

14. If $5 < m < 10$, then $m + 5 < 2m < m + 10 < m + 14 < m + 20 < m + 22$. The median is the average of the middle two: $[(m + 10) + (m + 14)] / 2 = \mathbf{m + 12}$.

A) $m + 12$ B) $m + 17$ C) $m + 21$ D) $(3m + 14)/2$

14. A

15. Since the lines determined by $4x + 10y = 12$ and $6x + 15y = 20$ are parallel, they will not intersect.

A) 0 B) 1 C) 2 D) 3

15. A

16. A large bundle has twice as many as a small bundle, so let's say a small bundle is s and a large bundle is $2s$. Jack has 5 large bundles and 7 individual stamps, so $5(2s) + 7 = 10s + 7$. Jill has 11 small bundles and 2 individual stamps, or $11s + 2$. Jack has the same number of stamps as Jill, so $10s + 7 = 11s + 2$, $5 = s$; each has **57** stamps.

A) 57 B) 67 C) 68 D) 78

16. A

17. Use the Remainder Theorem and let $x = 1$: $1^4 + 1^3 + 1^2 + 1 + 1 =$ **5**.

A) 1 B) 5 C) x D) $5x$

17. B

18. $(2★1) = 2(-1) = -2$, $(4★1) = 4(1) = 4$; $(-2★4) = -8(-14) =$ **112**.

A) -112 B) -54 C) 54 D) 112

18. D

19. Since $|3x + 5| = 2x$ cannot be negative, and no positive x works, **0**.

A) 0 B) 1 C) 2 D) 3

19. A

20. If $x^{400} = 9^{1000}$, $x^{400} = (3^2)^{1000} = 3^{2000} = (3^5)^{400}$, so $x = 3^5 =$ **243**.

A) 27 B) 81 C) 243 D) 729

20. C

21. If $(x^2 - 12)^2 = 169$, then either $x^2 - 12 = -13$, which has no real solutions, or $x^2 - 12 = 13$, which has **2** real solutions.

A) 1 B) 2 C) 3 D) 4

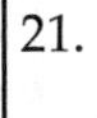

21. B

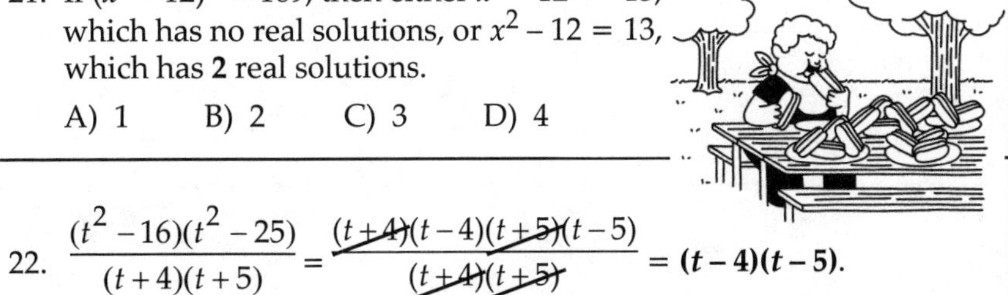

22. $\frac{(t^2 - 16)(t^2 - 25)}{(t + 4)(t + 5)} = \frac{\cancel{(t+4)}(t - 4)\cancel{(t+5)}(t - 5)}{\cancel{(t+4)}\cancel{(t+5)}} = \mathbf{(t - 4)(t - 5)}$.

A) $(t + 4)(t + 5)$ B) $(t - 4)(t - 5)$ C) $(t + 16)(t + 25)$ D) $(t - 16)(t - 25)$

22. B

Go on to the next page ➠ **A**

Solution	Answers
23. Since $\frac{12}{x}$ can equal any nonzero integer, there are **more than 12**. A) 5 B) 6 C) 12 D) more than 12	23. D
24. If $x^2 - 8x + b = 0$ has only one solution, it must be the square of a single binomial. Since $(x-4)^2 = x^2 - 8x + 16$, b must be **16**. A) 64 B) 32 C) 16 D) 8	24. C
25. Working together, 5 printers can print 640 pages in 4 mins. That equals 160 pages per min. for 5 printers, or 32 pages per min. for each printer. So 8 printers print 256 pages per min., and in **10** minutes they print 2560 pages. A) 8 B) 10 C) 12 D) 16	25. B
26. If n is an integer, then $n(n+1)(n+2)(n+3)$ has one multiple of 4 and another multiple of 2, and so is divisible by **8**. A) 8 B) 9 C) 10 D) 14	26. A
27. If b bags cost $m/10$ dollars, I can buy $b \times (d \div m/10) = \mathbf{10bd/m}$ bags. A) $\frac{10dm}{b}$ B) $\frac{10bd}{m}$ C) $\frac{10m}{bd}$ D) $\frac{bd}{10m}$	27. B
28. On Monday a jacket was marked down by 65% of its original price. Monday's discounted price was 35% of the original, x. Tuesday's decrease of 65% of 35% of x was \$91. Since $0.65 \times 0.35x = \$91$, $x = \$91 \div (0.65 \times 0.35) =$ **\$400**. A) \$400 B) \$450 C) \$500 D) \$550	28. A
29. $(3^3)^{2x+1} = 3^{6x+3} = (3^x)^6(3^3) = (2^6)(3^3) = 32a$; solving, $a = \mathbf{54}$. A) 18 B) 27 C) 36 D) 54	29. D
30. We know that $x + y = 30$, $y + z = 48$, and $x + z = 54$. Add the equations: $2x + 2y + 2z = 132$, so $x + y + z = 66$. The average is $66/3 = \mathbf{22}$. A) 22 B) 28 C) 36 D) 44	30. A

The end of the contest **A**

Visit our Website at http://www.mathleague.com

Answer Keys & Difficulty Ratings

2016-2017 through 2020-2021

ANSWERS, 2016-17 7th Grade Contest

1. B	8. A	15. D	22. B	29. C
2. D	9. B	16. C	23. A	30. B
3. C	10. D	17. A	24. B	31. C
4. B	11. D	18. B	25. B	32. B
5. A	12. D	19. A	26. B	33. A
6. C	13. C	20. B	27. C	34. C
7. A	14. C	21. D	28. A	35. D

RATE YOURSELF!!!

for the 2016-17 7th GRADE CONTEST

Score	Rating
34-35	Another Einstein
32-33	Mathematical Wizard
30-31	School Champion
27-29	Grade Level Champion
24-26	Best In The Class
21-23	Excellent Student
18-20	Good Student
15-17	Average Student
0-14	Better Luck Next Time

ANSWERS, 2017-18 7th Grade Contest

1. C	8. C	15. D	22. A	29. A
2. B	9. B	16. B	23. D	30. D
3. B	10. D	17. B	24. C	31. C
4. A	11. D	18. A	25. A	32. B
5. D	12. C	19. D	26. A	33. D
6. C	13. A	20. D	27. B	34. D
7. A	14. C	21. C	28. B	35. A

RATE YOURSELF!!!

for the 2017-18 7th GRADE CONTEST

Score	Rating
33-35	Another Einstein
30-32	Mathematical Wizard
26-29	School Champion
23-25	Grade Level Champion
20-22	Best In The Class
17-19	Excellent Student
14-16	Good Student
12-13	Average Student
0-11	Better Luck Next Time

ANSWERS, 2018-19 7th Grade Contest

1. C	8. B	15. C	22. C	29. C
2. D	9. C	16. B	23. B	30. D
3. A	10. D	17. C	24. B	31. A
4. A	11. B	18. A	25. D	32. C
5. D	12. C	19. B	26. B	33. D
6. B	13. D	20. A	27. D	34. C
7. C	14. B	21. D	28. B	35. D

RATE YOURSELF!!!

for the 2018-19 7th GRADE CONTEST

Score	Rating
32-35	Another Einstein
29-31	Mathematical Wizard
27-28	School Champion
25-27	Grade Level Champion
23-24	Best In The Class
21-22	Excellent Student
18-20	Good Student
15-17	Average Student
0-14	Better Luck Next Time

ANSWERS, 2019-20 7th Grade Contest

1. C	8. A	15. A	22. B	29. C
2. C	9. C	16. A	23. B	30. A
3. B	10. C	17. D	24. D	31. D
4. C	11. A	18. A	25. C	32. B
5. A	12. C	19. D	26. C	33. B
6. D	13. D	20. C	27. B	34. C
7. D	14. C	21. D	28. A	35. A

RATE YOURSELF!!!

for the 2019-20 7th GRADE CONTEST

Score	Rating
32-35	Another Einstein
28-31	Mathematical Wizard
25-27	School Champion
22-24	Grade Level Champion
19-21	Best In The Class
17-18	Excellent Student
15-16	Good Student
12-14	Average Student
0-11	Better Luck Next Time

ANSWERS, 2020-21 7th Grade Contest

1. D	8. A	15. D	22. C	29. D
2. A	9. D	16. C	23. C	30. C
3. C	10. D	17. D	24. A	31. B
4. A	11. C	18. D	25. D	32. B
5. B	12. B	19. B	26. A	33. D
6. A	13. C	20. C	27. A	34. B
7. C	14. B	21. A	28. D	35. C

RATE YOURSELF!!!

for the 2020-21 7th GRADE CONTEST

Score	Rating
32-35	Another Einstein
29-31	Mathematical Wizard
26-28	School Champion
23-25	Grade Level Champion
20-22	Best In The Class
18-19	Excellent Student
16-17	Good Student
14-15	Average Student
0-13	Better Luck Next Time

ANSWERS, 2016-17 8th Grade Contest

1. C	8. D	15. C	22. C	29. D
2. B	9. C	16. A	23. D	30. B
3. C	10. A	17. D	24. C	31. C
4. B	11. B	18. D	25. A	32. A
5. D	12. C	19. B	26. B	33. B
6. C	13. C	20. D	27. D	34. A
7. C	14. A	21. C	28. C	35. D

RATE YOURSELF!!!

for the 2016-17 8th GRADE CONTEST

Score	Rating
33-35	Another Einstein
30-32	Mathematical Wizard
27-29	School Champion
24-26	Grade Level Champion
21-23	Best In The Class
19-20	Excellent Student
16-18	Good Student
13-15	Average Student
0-12	Better Luck Next Time

ANSWERS, 2017-18 8th Grade Contest

1. D	8. C	15. D	22. D	29. B
2. B	9. C	16. B	23. A	30. A
3. D	10. B	17. A	24. A	31. A
4. B	11. A	18. A	25. C	32. A
5. D	12. D	19. B	26. D	33. C
6. B	13. A	20. B	27. C	34. B
7. C	14. B	21. D	28. A	35. C

RATE YOURSELF!!!

for the 2017-18 8th GRADE CONTEST

Score	Rating
34-35	Another Einstein
32-33	Mathematical Wizard
30-31	School Champion
27-29	Grade Level Champion
24-26	Best In The Class
22-23	Excellent Student
19-21	Good Student
16-18	Average Student
0-15	Better Luck Next Time

HOT STUFF

ANSWERS, 2018-19 8th Grade Contest

1. B	8. C	15. C	22. C	29. D
2. C	9. D	16. A	23. C	30. C
3. B	10. D	17. D	24. D	31. C
4. B	11. A	18. B	25. A	32. A
5. A	12. B	19. B	26. D	33. D
6. A	13. A	20. A	27. B	34. B
7. B	14. B	21. A	28. A	35. B

RATE YOURSELF!!!

for the 2018-19 8th GRADE CONTEST

Score	Rating
33-35	Another Einstein
31-32	Mathematical Wizard
29-30	School Champion
26-28	Grade Level Champion
24-25	Best In The Class
21-23	Excellent Student
19-20	Good Student
16-18	Average Student
0-15	Better Luck Next Time

ANSWERS, 2019-20 8th Grade Contest

1. D	8. B	15. B	22. B	29. D
2. A	9. C	16. B	23. C	30. B
3. B	10. A	17. D	24. D	31. D
4. B	11. B	18. A	25. C	32. D
5. A	12. C	19. D	26. C	33. A
6. C	13. C	20. A	27. A	34. C
7. C	14. A	21. B	28. B	35. D

RATE YOURSELF!!!

for the 2019-20 8th GRADE CONTEST

Score	Rating
32-35	Another Einstein
29-31	Mathematical Wizard
27-28	School Champion
24-26	Grade Level Champion
21-23	Best In The Class
18-20	Excellent Student
16-17	Good Student
13-15	Average Student
0-12	Better Luck Next Time

ANSWERS, 2020-21 8th Grade Contest

1. D	8. B	15. D	22. A	29. A
2. D	9. D	16. B	23. B	30. B
3. C	10. D	17. B	24. C	31. A
4. D	11. D	18. D	25. D	32. C
5. C	12. A	19. C	26. C	33. B
6. A	13. D	20. C	27. A	34. B
7. B	14. C	21. A	28. B	35. C

RATE YOURSELF!!!

for the 2020-21 8th GRADE CONTEST

Score	Rating
32-35	Another Einstein
30-31	Mathematical Wizard
28-29	School Champion
25-27	Grade Level Champion
22-24	Best In The Class
19-21	Excellent Student
16-18	Good Student
14-15	Average Student
0-13	Better Luck Next Time

ANSWERS, 2016-17 Algebra Course 1 Contest

1. B	7. A	13. A	19. C	25. D
2. C	8. A	14. C	20. B	26. A
3. D	9. C	15. A	21. A	27. B
4. B	10. C	16. C	22. C	28. D
5. A	11. B	17. D	23. A	29. C
6. D	12. C	18. D	24. D	30. B

RATE YOURSELF!!!

for the 2016-17 ALGEBRA COURSE 1 CONTEST

Score	Rating
29-30	Another Einstein
26-28	Mathematical Wizard
24-25	School Champion
21-23	Grade Level Champion
18-20	Best In The Class
15-17	Excellent Student
13-14	Good Student
10-12	Average Student
0-9	Better Luck Next Time

ANSWERS, 2017-18 Algebra Course 1 Contest

1. C	7. B	13. C	19. C	25. A
2. B	8. A	14. B	20. D	26. C
3. D	9. C	15. C	21. B	27. C
4. C	10. D	16. D	22. D	28. A
5. A	11. C	17. A	23. A	29. D
6. A	12. B	18. D	24. A	30. A

RATE YOURSELF!!!

for the 2017-18 ALGEBRA COURSE 1 CONTEST

Score	Rating
29-30	Another Einstein
27-28	Mathematical Wizard
24-26	School Champion
21-23	Grade Level Champion
19-20	Best In The Class
16-18	Excellent Student
12-15	Good Student
10-11	Average Student
0-9	Better Luck Next Time

ANSWERS, 2018-19 Algebra Course 1 Contest

1. C	7. B	13. B	19. A	25. C
2. B	8. C	14. A	20. B	26. B
3. D	9. B	15. D	21. D	27. B
4. A	10. B	16. D	22. C	28. B
5. B	11. B	17. B	23. A	29. A
6. C	12. D	18. B	24. D	30. D

RATE YOURSELF!!!

for the 2018-19 ALGEBRA COURSE 1 CONTEST

Score	Rating
28-30	Another Einstein
25-27	Mathematical Wizard
23-24	School Champion
20-22	Grade Level Champion
17-19	Best In The Class
13-16	Excellent Student
11-12	Good Student
9-10	Average Student
0-8	Better Luck Next Time

ANSWERS, 2019-20 Algebra Course 1 Contest

1. A	7. A	13. D	19. C	25. A
2. B	8. B	14. C	20. A	26. B
3. C	9. B	15. A	21. B	27. B
4. C	10. A	16. C	22. A	28. C
5. D	11. C	17. B	23. D	29. D
6. B	12. A	18. D	24. D	30. D

RATE YOURSELF!!!

for the 2019-20 ALGEBRA COURSE 1 CONTEST

Score	Rating
28-30	Another Einstein
25-27	Mathematical Wizard
22-24	School Champion
19-21	Grade Level Champion
16-18	Best In The Class
13-15	Excellent Student
11-12	Good Student
9-10	Average Student
0-8	Better Luck Next Time

ANSWERS, 2020-21 Algebra Course 1 Contest

1. A	7. D	13. C	19. A	25. B
2. C	8. B	14. A	20. C	26. A
3. D	9. A	15. A	21. B	27. B
4. C	10. B	16. A	22. B	28. A
5. B	11. C	17. B	23. D	29. D
6. D	12. C	18. D	24. C	30. A

RATE YOURSELF!!!

for the 2020-21 ALGEBRA COURSE 1 CONTEST

Score	Rating
29-30	Another Einstein
26-28	Mathematical Wizard
24-25	School Champion
21-23	Grade Level Champion
18-20	Best In The Class
15-17	Excellent Student
12-14	Good Student
10-11	Average Student
0-9	Better Luck Next Time

WE MADE IT!